Tracks along the Ditch, Relationships between the C&O Canal & the Railroads

Patrick H. Stakem

© 1998
4th Edition, 2018

"Success to the Navigation of the Potomac!"
G. Washington

Contents

Author..6
Acknowledgments...6
Introduction...6
Brief History of the C&O Canal...14
Chronology..23
Connection with other canals and other rivers................................26
   River Locks..27
   The Patowmack Company..29
   Alexandria Canal and the Aqueduct Bridge................................31
   Annapolis & Potomac Canal Company..32
   Chesapeake & Delaware Canal..32
   Frederick County Canal Company...32
   Maryland Cross-Cut Canal...33
   Pennsylvania Canal 34
   Potowmack Canal...35
   Shenandoah Canal..35
   Washington City Canal..36
   The Potomac River...37
      The South Branch of the Potomac River..................................37
      The North Branch of the Potomac...37
      Will's Creek..38
   The Ohio River...38
Railroads..40
   Amtrak 41
   Baltimore & Ohio...41
   Billmeyer Lumber Company Railroad..43
   Bloomington and Fairfax Railroad...43
   Chesapeake & Ohio Railroad (Chessie System)..........................44
   Conrail..44
   CSX...45
   Cumberland & Pennsylvania Railroad...46
   Frederick and Pennsylvania Railroad...49
   George's Creek & Cumberland...50
   George's Creek Rail Road..52
   Georgetown Barge, Dock, Elevator, & Railway Company........54

Green Ridge Railroad...54
MARC..55
Norfolk Southern...62
Norfolk & Western RR..63
Pennsylvania Railroad in Maryland..63
Piedmont & Cumberland Railroad 64
Pittsburgh and Connellsville Railroad......................................65
Potomac and Piedmont Coal and Railroad Company...............66
Salisbury and Baltimore Railroad and Coal Company.............66
Shenandoah Valley Railroad...67
South Branch Valley RR..67
Twin Mountain & Potomac...67
Washington & Cumberland Railroad..68
Washington & Old Dominion Railroad....................................70
Washington & Western Maryland Railroad.............................71
Western Maryland Railroad..71
Western Maryland Scenic Railroad..73
West Virginia Central & Pittsburgh Railroad...........................75
Williamsport, Nessle, and Martinsburg....................................77
Winchester and Potomac Railroad..77
Winchester & Western Railroad..78
East to west; Tidewater to the mountains.....................................80
  Georgetown (MP 0.0)...80
  Point of Rocks (MP 48)..82
  Brunswick (MP 53-54)..83
  Harpers Ferry (MP 60.7-61)..83
  Antietam (mp 69)...85
  Shepherdstown (MP 72.2)...85
  Williamsport (MP 99)..86
  Green Spring, MD (MP 110)...86
  Big Pool (MP 113)...87
  Hancock (MP 124)...87
  Little Orleans (MP 131)...88
  Paw Paw (MP 156)...88
  Mertens Sawmill; Green Ridge Railroad (MP 160).................89
  Town Creek Aqueduct (MP 162)..89
  Oldtown, MD (MP 166.7-167)..89

Patterson Creek (MP 173.5)..90
Western Terminus – Cumberland, MD, Milepost 184.5.............90
The National Road Era...93
Boatyards..94
   Mertens Boatyard...96
Railroad Connections..96
   Boat wharfs...97
   Railroad Facilities...99
   WVaC&P 100
   GC&C..100
   Consolidation Coal Company...100
   Cumberland Coal & Iron Company...101
   Cumberland & Pennsylvania Railroad....................................102
Passenger traffic from Cumberland..104
   Maryland Mining Company Rail Road....................................104
   George's Creek and Cumberland Railroad..............................105
   Cumberland & Pennsylvania Railroad 105
   Baltimore & Ohio Railroad...106
   West Virginia Central and Pittsburg Railroad........................106
   Western Maryland Railway..107
   Mount Savage Rail Road...107
   Pennsylvania Railroad in Maryland...107
   The trolley...108
   Amtrak..109
Relative Economics of Transportation Options............................110
   Horses versus steam..112
   Horse versus Mule...114
   Conestoga wagons...115
   Stage coaches...116
   Flatboats..117
   Canal boats...117
Intermodal wins..118
Traffic Analysis..121
   What cargo moved on the canal?..121
   Downstream freight..123
   Flour shipments...126
   Coal...127

What happened to the coal from Cumberland?..........................129
Enabled Industries..........................131
West of Cumberland..........................136
   Canal Extension, option 1..........................138
   Canal extension, option 2..........................143
   Westernport, MD and Piedmont WV..........................145
   The B&O's 17-Mile grade..........................146
   Pittsburgh, PA, and the Pennsylvania Canal System..........................146
   Western Maryland Rail Trail..........................147
For Further Reading..........................149
   Places..........................155
   Lonaconing..........................158
   Mt. Savage, MD..........................159
   Washington, DC..........................160
   Amtrak..........................160
   Chessie and CSXT..........................162
   Conrail..........................163
   CSX..........................163
   Cumberland Valley Railroad..........................163
   Historic Railroad Lines..........................163
   MARC..........................166
   Traction..........................166
   Western Maryland Railroad..........................167
   Western Maryland Scenic Railroad..........................168
   T. H. Paul..........................168
   National Road..........................169
   Misc topics..........................170
Notes..........................173
Videos..........................173
Websites..........................174
If you enjoyed this book, you may enjoy some of my other titles...........................175

Author

The author grew up in Cumberland, MD, across the street from the location of Fort Cumberland. His material grandparents lived in Eckhart, and his Grandfather, "Jake" Ward was an engineer for the Cumberland & Pennsylvania Railroad.

Mr. Stakem is a member of the Council of the Alleghenies, as well as the Western Maryland Chapter of the National Railway Historical Society, the Mount Savage Historical Society, the Westernport Heritage Society, and the C&O Canal Association. He currently resides in Laurel, MD. He has published numerous articles and books on the railroads, canal, and industrial and transportation heritage of Western Maryland. He is associated with Loyola University in Maryland and the Johns Hopkins University.

Acknowledgments

Special thanks to: Bob Shives, Photo Archivist of WMRHS; the archives of the Western Maryland Railway Historical Society, Union Bridge, MD; the archives of the Western Maryland Chapter of the National Railway Historical Society, Cumberland, MD; the C&O Canal Historical Society; the C&O Canal Visitors Centers, Cumberland, Hancock, Great Falls, Georgetown, and Williamsport; Kate Mulligan; Dr. Karen Gray of the C&O Canal Library in Hagerstown; Al Feldstein, Frank Tosh, Arlington, Va., Traction/Trolley Expert Extraordinaire; The Preservation Society of Allegany County, Cumberland, MD; The Allegany County Library System; The Appalachian Collection at Allegany College of Maryland; Cumberland, MD; The Mount Savage Historical Society, Rita Knox, NPS Ranger at Cumberland, MD; Gary Petrichick, Jim Dilts, Bob Bantz, William Bauman, Eileen Carlton, and many others.

Introduction

This book explores the interaction of the various transportation technologies, driven by the profit motives and political situations of the time. The interaction of the Canal and the Railroads with the National Road is also touched upon.

This book presents a guide to the many railroads that touched the path of the Chesapeake & Ohio Canal. It is intended for the person interested in both canal and early railroad history. The Baltimore and Ohio (B&O) Railroad and the Chesapeake and Ohio (C&O) Canal projects broke ground on same day, July 4, 1828, with the same goal, reaching the rich grain traffic on the Ohio River. The canal winds 184.5 miles from Georgetown near Washington, D. C. to Cumberland, Maryland. The B&O Railroad stretches 179 miles between Baltimore and Cumberland. The railroad reached its goal in 1853, but canal construction stalled at Cumberland. The canal company is out of business, and the old canal towpath today is used by hikers and bikers. The canal is preserved as a National Park, thanks to the efforts of Supreme Court Justice William O. Douglas and many others. The B&O route is still in daily use for both

passenger trains and heavy freight. The National Road still exists, but has mostly been superseded by an Interstate. Both canal and railroad, along with the National Road, formed part of the important Trans-Appalachian transportation system, and all were, in 19th century terms, Methods of Internal Communications.

The year 2011 marked the 200th anniversary of the initiation of the National Road project by President Jefferson, the first of a series of internal improvements for the country. It ran west from Cumberland. Before the railroad and the canal were built, freight was moved by wagon, and people moved by stagecoach. Initially, the stagecoach was used to complete the path to Wheeling, VA (now WV) on the Ohio River. Later, the stagecoach industry was pushed farther west, and phased out when the railroad reached the Ohio River. Both the railroad and the canal would be superseded by more and better roads, cars, and trucks. Time and technology marches on.

The Baltimore to Cumberland section of the road was designated the Baltimore National Pike. It combined several existing turnpikes such as the Baltimore and Fredericktowne Pike, the Hagerstown and Boonsboro Turnpike, and the Cumberland Turnpike. Local banks financed the pike, which became known as the Bank Road.

When it was constructed, the canal was based on tried and proven technology, and had an extensive history in Europe. The railroad was new, untried technology. At its inception, it was not clear if locomotives or horses would be used on the railroad. Horses were used initially. The B&O Railroad's inclined planes at Mt. Airy, Maryland, were designed so that stationary steam engines could pull the cars up the hills. This scheme was used for canal boats on the Allegheny Portage Railroad further north in Pennsylvania. The B&O's locomotives proved capable of handling the mountains west of Baltimore by themselves, and the horses were out of a railroad job.

Interestingly, the C&O Canal did not go from the Chesapeake not to the Ohio. Perhaps we should call it the Georgetown and Cumberland Canal? Begun in 1828, it took 22 years to reach Cumberland. Over its 184.5 miles, it rises 610 feet from tidewater, and uses 74 locks.

The C&O Canal owed its existence to the much earlier Patowmack Canal Project, championed by George Washington. He saw the value in transportation routes across the Appalachians, and hoped to utilize the Potomac River, with a portage in the western part of the state, to reach the Ohio River. The Ohio was then and still is a major North-South transportation system west of the Appalachians, and reaching to the Gulf of Mexico. Washington was very familiar with the territory, having traveled across it many times on horseback and canoe downstream from Fort Cumberland. His canoe trip in 1754 after his defeat by the French at Fort Necessity convinced him of the value of the river as a transportation mode.

The first problem with the navigation of the Potomac is Great

Falls, with a drop of 80 feet in a mile. Here, the major project was the skirting canal on the Virginia side. Large sections of this remain, and are managed by the National Park Service. In 1784, the States' Assemblies of Virginia and Maryland adopted Washington's plan to form a company to exploit the Potomac for transportation. Washington was the first President of the Company, having resigned his military commission after the Revolution. Meeting at his home, Virginia and Maryland Delegates drew up the Mount Vernon Compact, which provided for free trade on the Potomac. Construction was started in 1785 and continued for 17 years. Meanwhile, the new Federal City of Washington was built downstream of Great Falls, and just upstream from Washington's plantation at Mount Vernon.

Soon after opening, thousands of boats used the locks to carry flour, whiskey, tobacco and iron downstream, and manufactured products upstream. Downstream trips required 3-5 days from Cumberland. Upstream was much more difficult, requiring 10-12 days by poling against the current. Additional skirting canals were needed at Little Falls and Harper's Ferry. And then, the Potomac usually had either too much water, or not enough, to support transportation. Rafts or keel boats of 20 tons capacity could make the downstream trip for only about an average of one month in a year. Washington died in 1799, two years before the canal opened, and did not get to see his dream realized. The Patowmack Canal Project was a vast improvement, but was not the success that was hoped for.

The construction costs were excessive, and the maintenance costs exceeded the revenue. Lessons learned were that the Potomac is not a good artery of transportation. A completely separate canal would be needed. The Patowmack Company surrendered its charter in 1828 to the Chesapeake and Ohio Canal Company, which also assumed the assets and liabilities of the earlier company.

The C&O Canal faced its own difficulties in raising money, paying off construction costs and maintenance expenses, and getting enough labor to get the canal built. The choke point at Point of Rocks, Maryland, first brought the canal and the railroad into legal contention over access, and they were already competing for the labor and funds for construction.

The coal trade was being driven by large companies, principally New York interests. The demand for coal was great, and the supplies in Western Maryland were seemingly limitless. (Coal is still mined there). The canal and the railroad were focused on the grain trade and were not ready for the coal traffic. The demands of the various local short line railroads, mining companies, and iron works in Western Maryland changed the focus of the long-haul freight providers. Coal went out, first by wagons and flat bottom boats, but later by canal and railroad, initially to the east, but later west to the steel works in Pittsburgh.

The origin of the Western Maryland coal business began in the early 19th century when a 14-foot thick seam of bituminous coal referred to as The Big Vein was discovered in the George's Creek Valley. This coal region became famous during the industrial revolution in the 19th century for its clean-burning low sulfur content that made it ideal for powering ocean steamers, riverboats, locomotives, steam mills, and machines shops. However, coal production did not really become important until after the B&O Railroad reached Cumberland in 1842. In 1850, the opening of the Chesapeake & Ohio Canal from Georgetown to Cumberland provided another route for coal shipments. The cost of transportation dropped, and the coal market took off.

By 1850, some thirty coal companies were mining the George's Creek Valley, producing over 60 million tons of coal between 1854 and1891. The Consolidation Coal Company, Consol, was established in 1864 and headquartered in the city of Cumberland, MD for its first 85 years (1864-1945).

Western Maryland's coal production reached about 1 million tons in 1865, and exceeded 4 million tons by the turn of the century. It reached an all-time high of about 6 million short tons in 1907. A small amount of the coal production in the early 1900s was premium blacksmithing coal that was specially processed and delivered in boxcars to customers throughout the United States and Canada.

At Cumberland, the C&O Canal reached its western terminus. There, numerous short-line railroads fed coal to the canal boats. Boat yards provided new construction and repairs, and provisioners equipped families for the week's trip down to Georgetown. Cement was also shipped from Cumberland for canal construction from quarries along Will's Creek to the west of the Market Street Bridge. Iron fixtures from the Lonaconing furnace were used to lock stones together in the canal locks. Manufactured goods came up the canal to Cumberland, although the railroad became the main avenue of shipment for freight and for passenger service. A minor passenger service was provided on the canal, but you couldn't be in a hurry. Packet boats carried the mail.

The canal and railroad provided an intermodal transportation system. Coal reached the Cumberland terminus by railroad, where it was loaded into the canal boats. Before the canal reached Cumberland, the B&O railroad hauled coal downstream to Williamsport or Hancock, where it was transferred into canal boats for the trip south to Georgetown. Near Williamsport, a short line railroad hauled coal from canal boats up into Pennsylvania. And, speaking of hedging your bets, the Roundtop Cement facility had both canal and rail access for shipment of their products.

The canal wasn't built to serve the coal fields of Cumberland. It was headed to the Ohio River to tap into the rich grain trade. Had the canal proceeded west from Cumberland, which way would it have gone? There was an easy water level route to the aptly named Westernport, and a possible connection to the George's Creek coal fields. This was also the path of the original B&O main stem. Also, one could venture west from Cumberland along Will's Creek, through the Narrows and then north to Pennsylvania. Later, the B&O would go this way also, using the Sand Patch grade. Both routes involved steep grades and tunnels through the mountains.

Railroad history was being made along the way, not only on the eastern side of the mountains, as the B&O struggled to invent American Railroading, but also in the high technology iron manufacturing center of Mt. Savage, some 10 miles from Cumberland. Here, the technologists did what was said to be impossible; they rolled the first iron rails in the United States, as good as the British product, and cheaper. This disruptive event was to change the world economy. The British stranglehold on world-wide rail production was broken. The B&O had built west with imported rail, which came into the Port of Baltimore via sailing ship. The Mt. Savage Rail Road built their line using their own product. Now, the B&O could buy a US product. In 1846, the B&O contracted for 15 miles of Mount Savage rail, nearly 675 tons of the 51 pounds-per-yard products then produced. Times were changing.

We will discuss significant railroad locations along the canal. Distances generally follow Hahn's towpath guide, 25th anniversary edition. The canal can be hiked from end to end, or can provide a series of weekend trips. You can ride canal boats in Georgetown or at Great Falls and, possibly soon, at Cumberland. At the moment, a key railroad bridge blocks the extension of the re-watered portion of the canal. Imaging that – a canal-railroad controversy over right-of-way. The Great Allegany Passage allows a through hike or bike from Pittsburgh to Washington, DC. A steam locomotive pulls a tourist train between Cumberland and Frostburg. The National Road heads westward not too far from the train station and the canal terminus. Its successor, Interstate-68 bypasses all its predecessors in Cumberland on an overhead bridge.

A note on one confusing issue about what state various towns are located in is worth mentioning. With the initiation of the Civil War, Virginia seceded from the Union. In 1863, the western part of the State of Virginia, from Harpers Ferry westward, seceded from Virginia. It was admitted to the Union as the new state of West Virginia. So, before 1863, Harper's Ferry, Shepardstown, Ridgeley, Keyser, Piedmont, and other towns were in Virginia. After that, they were part of West Virginia.

Notes to the second edition
There are some updates and formatting changes to this edition to enhance clarity, and some corrections and edits. Note that certain material is repeated between sections, because of the relevance to both.

Brief History of the C&O Canal

It was on the Fourth of July in 1828 that both the Baltimore & Ohio (B&O) Railroad and the Chesapeake & Ohio (C&O) Canal broke ground for building westward from the eastern seaboard at Baltimore and Georgetown, respectively, to the Ohio River. The geography dictated an intermediate goal of Cumberland, Maryland. The B&O reached Cumberland in 1842; the Canal came in 1850. The canal represented proven European technology, and the railroad was a more speculative venture. Although grain was intended to be the major cargo, the black gold (coal) from the rich fields around Eckhart and the George's Creek quickly became the dominant cargo for both the canal and the railroad. The State of Maryland invested heavily in both ventures.

The Allegany coal trade kicked off in December 1843 when coal was hauled by wagon on the National Road from Eckhart to Cumberland. There, it was loaded onto the B&O railroad. The coal was transported by rail to dam #6 on the C&O canal west of Hancock, where it was transferred to canal boats for the trip to Georgetown. There, it was loaded on sailing ships for the journey up to New York. And, New York was glad to get it. But there had to be a better way.

The eight-year head start (from 1842 to 1850) was enough to provide the freight monopoly to the railroad. The canal, stalled at Williamsport, proved a major customer for the B&O, however. Trans-shipment of coal was done for the Washington / Georgetown / Alexandria / New York markets. In a remarkable cooperation, the B&O hauled coal from Cumberland to Williamsport, where it was loaded onto canal boats for the remainder of the trip. This practice continued until the Canal reached Cumberland. Then, the canal boats could be loaded with coal directly. By that time, the railroad had pushed west, nearly to Piedmont, Virginia (later, West Virginia), at the base of Backbone Mountain. This allowed more coal from the George's Creek Valley and West Virginia to be tapped by the railroad.

Westernport and Piedmont, sister cities across the Potomac River, became a logical target for connection of rail and canal with the George's Creek region. One of the two proposed paths for the canal westward from Cumberland to the Ohio River would have passed through Westernport. Unfortunately, the Canal Company ran out of money, and the canal ended at Cumberland. Before then, the George's Creek Coal & Iron Company had connected the iron furnace and mines at Lonaconing via rail line to Westernport and Piedmont. They produced wrought iron dowels to lock the stones used in the lock construction together. The B&O reached Piedmont in July 1851. The George's Creek Coal & Iron Company built their line from Piedmont to Lonaconing in 1852. That line was operated by the George's Creek Rail Road, and was acquired in 1863 by the Cumberland & Pennsylvania (C&P) Railroad.

After 1850, shipping coal by canal was cheaper than by the railroad, and it provided the most direct, though not the quickest path to the Washington, D.C. area. Coal is a time-insensitive cargo; it doesn't deteriorate during transport. Trains from Cumberland to Baltimore typically took 10 hours, where a canal boat trip from Cumberland to Georgetown was 5 days. Coal flowed into Cumberland from the George's Creek and Eckhart mines over a series of short line railroads, mostly consolidated into the Cumberland & Pennsylvania Railroad by 1870. To get to the canal loading wharves, the C&P had to cross B&O tracks. The B&O extracted a toll for this passage.

The original coal wharves allowed bottom dumping of a coal car directly into a canal boat which was a major improvement over the hand-shoveling method. Originally the facility allowed only for one loaded coal car at a time, pulled by horses. The structure was not strong enough to support the weight of the locomotives. Later, Consolidation Coal would build a large concrete structure for canal boat loading. A typical canal boat held 120 tons of bulk cargo. Railroad cars of the time would hold 10 tons, so a dozen cars would be loaded into one boat. Modern coal cars are typically of

100 or 120 ton's capacity. They travel in 100 car unit trains, 10,000 tons of coal at a time.

Although they are often portrayed as antagonists, the canal and the railroads often cooperated in moving the black gold of the Cumberland region to the insatiable markets of the Eastern seaboard. Before the canal was completed to Cumberland, the Baltimore & Ohio Railroad took coal from Cumberland to Williamsport for trans-loading onto canal boats. The Canal wound up being controlled by the B&O, who defended its existence, because it did not want a rival to buy the right-of-way. The hard part of construction was done; it would be a simple matter to fill in the canal prism and run a railroad to Washington. It would be subject to the same flooding conditions of the Potomac that the canal was, but it would provide a major advantage to competing rail lines. We'll discuss this in some detail later.

In the early 19th century, the Western Maryland region had vast natural resources of agricultural materials, coal, iron, fire clay, and wood. The problem was that there was no practical way to get these materials to the markets along the seaboard. This situation had been recognized by a surveyor in the employ of Lord Fairfax, George Washington. He not only went on to become not only President of a new Republic, but President of a canal company, formed to improve navigation on the Potomac. The Patowmack Canal Company was a direct predecessor of the C&O Canal. However, the first transportation project that touched the western part of the state was the National (or, Cumberland) Road, begun under President Jefferson. This followed the Native American trails, and parts of Braddock's military road, itself built along the trade-route trails blazed by the Ohio Company in the 18th Century. The Native American trails generally followed the paths through the forests and over the mountains blazed by countless migrations of bison, elk, and other large herd animals over millennia. Migrating herds of animals, with their young, will choose the easiest path. To them we owe the paths of our railroads and highways.

From the Western Maryland Station in Cumberland, MD, where the canal terminus is located, one can see simultaneously the railroad, the canal, and the start of the National Road. Interstate-68, successor to the National Road, passes overhead on a bridge. All of the 19th Century transportation modes come together in one place here.

In 1828, and through the 1830's, passengers walked, rode horses, or took the various stage lines over the National Road. Before the road was completed, pack horses were used. Freight moved by heavy wagon, or down the Potomac in flat bottom boats at limited times of the year. There was no practical way to return the boats upriver -- they were sold off for their wood in Georgetown and Alexandria. When the B&O started building the railroad, it was not at all clear what was the best motive power -- horses or steam engines. This would be decided later in favor of the steam engine at the Rainhill Trials, in England. The B&O started out as a horse-drawn system.

In the late 1820's, Allegany County represented the unlimited frontier, and the canal and the railroad were coming. This point was not lost on the early entrepreneurs of the region, as well as industrialists from New York and New England, and merchants from Baltimore. Both railroad and canal would pass through the town on their way west, and both would provide an avenue for transportation of goods -- and ways to make a profit, if not a fortune. New York and European money poured into the region.

The earliest railroads in Allegany County were constructed by and for the coal mining and iron producing companies. They were built of necessity to move products to market. From the extraction or production site to the railhead or the canal terminus, the railroads included the Eckhart, built in 1846 by the Maryland Mining Company, the George's Creek, built in 1853 by the George's Creek Coal & Iron Co, and the Mount Savage Rail Road, built in 1845 by the Mount Savage Coal & Iron Company.

Christian Detmold, the operator of the Iron Furnace in Lonaconing was responsible for the construction of an early tram road in 1847 from Lonaconing to Clarysville, to connect with the Eckhart Railroad. This was an attempt to provide transportation for iron goods from Lonaconing. The tram road was horse powered, and used wagons on wooden rails, covered with strap iron. The George's Creek Railroad later connected Lonaconing with the B&O railhead at Piedmont, which was also a target for the extended canal. Originally intended to transport finished iron from the furnace at Lonaconing, the rail line quickly switched over to being a coal carrier, from the Big Vein along the George's Creek.

The Maryland & New York Iron & Coal Company's Mt. Savage Railroad had been in operation from Mt. Savage to Cumberland since 1845. Much of the initial rolling stock and motive power was provided by the B&O. Ross Winans of Baltimore and other builders provided equipment to the mining railroads. Rail for the Mount Savage Rail Road was produced locally at Mt. Savage, breaking the English monopoly on rail manufacturing. The B&O purchased rail from the Mt. Savage Works to update their main line near Harper's Ferry.

In 1850, the year the Cumberland & Pennsylvania (C&P) Railroad was chartered, the George's Creek Coal & Iron Co. furnace at Lonaconing was in intermittent service, and the blast furnaces built by the Maryland & New York Company in Mt. Savage were in full operation. The first iron rail had been rolled in 1844, but the company had failed in 1848. The assets of the company were quickly picked up at sheriff's sale by another entity. Also in 1850, the canal reached Cumberland. That year, there were three shortline railroads operating in Allegany County, and coal could be transferred to canal boats in the Potomac or Will's Creek.

In July 1851, the B&O reached Piedmont, Virginia. The George's Creek Coal & Iron Company built their rail line from Piedmont north along the George's Creek to Lonaconing in 1852. That line was acquired in 1863 by the C&P. The shops and engine house at Lonaconing were used until 1867.

The Coal and iron Company had previous surveyed three potential lines to get their product from Lonaconing to Cumberland, and went so far as to buy land for a wharf facility in 1839. But the canal still was 50 miles east of Cumberland.

The Mount Savage Iron Company extended their rail line westward 3 miles from Mount Savage north to Borden Yard in 1851. The C&P Railroad, now part of Consolidation Coal, acquired the properties and equipment of the three roads discussed over the period 1850-1870. The Mount Savage Railroad was acquired in 1854 and the George's Creek in 1863. The Eckhart railroad operated concurrently with the C&P for twenty years, until finally being acquired in 1870.

All of the original lines were now part of the C&P, and if you wanted to move coal, you talked to them. The C&P was owned by the Consolidation Coal Company, as were most, but not all, of the mining companies. Other mining companies had issues with this monopoly. Several alternatives were discussed, including connection with lines other than the B&O, and construction of competitors to the C&P. The Pennsylvania Railroad was always happy to take traffic away from the rival B&O.

The Civil War affected both the canal and the railroad. The B&O line was in Virginia from Harpers Ferry to near Cumberland. The vulnerabilities and importance of the canal and the railroads became readily apparent to both sides of the conflict. The union side had a lot of man power tied up in defending the transportation arteries. At the end of the war, the value of rapid and dependable transportation to warfare was established beyond a doubt. Many of the best military engineers went into the railroad building business. A railroad soon stretched across the continent. Later, a canal would also link the Atlantic and Pacific, but even at the narrowest point in Panama, it was a long and expensive project.

Further up the Potomac from Cumberland, Senator Henry Gassaway Davis from West Virginia sought a way to bypass the monopoly by the B&O Railroad's transportation of coal. He had his West Virginia Central and Pittsburgh Railroad in West Virginia. He built a rail line up to Cumberland (the Piedmont and Cumberland Railroad) along the opposite side of the Potomac than the B&O was using, and convinced the Pennsylvania Railroad to build down to Cumberland as well. The connection between the lines was at Cumberland. The George's Creek and Cumberland Railroad was built to haul coal out of the George's Creek, and pass it to the Pennsylvania Railroad as well. This was all not without controversy.

Davis kept an eye on canal finances, looking for a cheap acquisition. He didn't want a canal; he wanted the much more useful right-of-way from Cumberland to Georgetown. Buy it at fire sale prices, fill it in, pave it over, lay track. The hard part of the construction was done. The B&O, holder of a lot of canal debt, blocked the scheme. However, the Maryland Board of Public Works wanted its construction bond money back. In 1895, it advertised for bids on the millions of dollars of canal bonds it held. It received bids from Elkins, his business partner Kerens, and the Washington and Cumberland Railroad.

The Washington and Cumberland Railroad had been incorporated in 1890. It was protected by a stipulation that it couldn't be owned by a railroad company owning competing track, meaning the B&O. The Maryland Governor supported the new Company. The DC City Council granted it access to the City. Some of the incorporators of the W&C included Enoch Pratt, of Baltimore, and John Hambleton, a Baltimore Banker and business partner of Davis. Interestingly, the take-over scheme was supported by the ex-President of the Canal, Arthur Gorman, also a partner of Davis. It looked like the canal was going to become a rail line. The W&C went so far as to have Charles Latrobe do a survey. Charles was the son of famed B&O civil engineer Benjamin Latrobe.

The W&C would also branch off at Williamsport to get to Baltimore, a plan supported by the Western Maryland Railway. The West Virginia coal would flow to the seaboard via a path not involving the B&O.

It was a long, involved, messy story, but the B&O prevailed. It controlled the receivership of the Canal. That receivership contracted with a shadow corporation of the B&O, the C&O Transportation Company, to operate the Canal. The B&O had emerged from bankruptcy, and didn't need another rival. It also didn't need a canal, but it couldn't afford to let anyone else have it either. Davis was not happy, but a cash payment and an agreement on favorable rates eased the pain somewhat. The State was paid for its bonds, not as much as it wanted, but it was now happily out of the canal business as well. The C&O Transportation Company operated the canal until it was sold to the government.

In the endgame, a Gould-led syndicate bought Davis' railroads as part of a planned intercontinental railroad scheme. The Western Maryland Railway was to be a key part. Financing fell through, and the syndicate dissolved. The Western Maryland Railway wound up in bankruptcy, but emerged as the Western Maryland *Railroad*, including most of the ex-Davis lines and the Western Maryland coal shortlines as well.

The big floods of 1924 did a lot of damage to the railroad bridges and lines, and dealt the death blow to the canal. It never recovered from the damage. The railroads were up and running shortly after the water level went down. The Canal went into bankruptcy, and the receivers suspended navigation. The lower few miles were maintained to continue water rights in Georgetown for Industry. It was not legally abandoned, but with maintenance suspended, it was unlikely to ever be used again.

Later, the B&O took over the Western Maryland, and closed down most of the redundant trackage. Then the B&O became part of the

Chessie System, and later CSX Corporation, which still operates along the canal today.

Chronology

1749 - Christopher Gist sees commercial potential at junction of Will's Creek and the Potomac.
1751 - Road by Ohio Company from site of Cumberland to Forks of the Ohio (Pittsburgh).
1755 - Braddock expedition against Fort Duquesne from Fort Cumberland.
1764 - Mason-Dixon survey of border between Maryland and Pennsylvania.
1775-1783 George Washington as Commander in Chief of the Continental Army.
1784 - Washington Expedition (one of many) to "the West"
1785 - Patowmack Company chartered with G. Washington as President
1787 - Demonstration of Rumsey's steamboat on the Potomac near Shepherdstown, WV.
1789-1797 George Washington serves as the first President of the United States.
1806 - National Road Project authorized by President Thomas Jefferson.
1811 - National Road construction began.
1818 - National Road reaches Ohio River at Wheeling, VA (later, WV).
1823 - Patowmack Company resolved to surrender its charter to a new company.
1824 - C&O Canal Company chartered; Patowmack Company surveys of canal west of Cumberland.
1828 - July 4, Ground broken for the C & O Canal; B&O Railroad starts out from Baltimore.
1829 - Rainhill Trails of Locomotives in England shows feasibility of use.
1831 - Canal navigation between Georgetown and Little Falls B&O Railroad reaches Frederick, MD.
1832 - B&O Railroad reaches Point of Rocks, MD.
1833 – C&O Canal to Harpers Ferry completed (Mile 60).
1837 - Blast furnace operating in Lonaconing, MD.

1839 – C&O Canal to near Hancock, MD completed (Mile 134).
1842 – C&O Canal construction suspended
1842 – B & O Railroad reaches Cumberland, MD. Mount Savage Rail Road meets them.
1847 – C&O Canal construction resumes
1850 – C&O Canal completed as far as Cumberland, MD.
1851 - B&O Railroad reaches Piedmont, VA.
1852 - George's Creek Rail Road reaches Piedmont from Lonaconing.
1853 - B&O Railroad reaches its goal of the Ohio River at Wheeling, VA.
1863 – West Virginia secedes from Virginia, is accepted into the Union.
1870 - Cumberland & Pennsylvania Co. absorbs all the mining railroads in Allegany County, MD.
1889 - Major devastating flood causing Canal to go into receivership; acquired by B & O Railroad
1892 – C&O Canal repaired and put back in operation.
1902 - Canal Towage Company established. End of owner-operated boats on the canal.
1924 - First major flood in 35 years. The ruined Canal closed down permanently.
1938 – Sept. 23; C&O Canal becomes a unit of the National Park Service.
1939 – C&O Canal dedicated as a public park.
1950 - Proposal for parkway to Cumberland along Potomac using the canal.
1954 - Justice William O. Douglas writes letter to editor of Washington post, for a hike on the Canal.
1954 - Hike from Lock 72 to D.C.; editors concede the canal should be preserved.
1971 - President Nixon signed an act to establish the C&O Canal National Historical Park.
1977 - C&O Canal was dedicated to Justice Douglas.
1988 - Scenic Railroad begins operations in Cumberland with steam locomotive.
1991 - Interstate-68 opened through Cumberland; successor to

National Road.
1993 - CanalPlace in Cumberland established by the Maryland General Assembly.
2006 - Great Allegany Passage opened from Pittsburgh to Washington, DC via Cumberland.

Connection with other canals and other rivers

Although we are focusing on the interconnections, intersections, and interactions of the C&O Canal and the railroads, we also need to consider the connection of the canal with other canals and rivers. The C&O Canal was built along the Maryland side of the Potomac River, which was previously used to transport goods. The river was fickle - there was too much water in the spring, and not enough in the fall. There was still river traffic after the canal became operational; river travel was free. River locks provided access to the canal, particularly for boats from the Virginia side. The rivers that intercepted the south side of the Potomac from Virginia, Patterson's Creek, the South Branch of the Potomac, the Shenandoah, the Cacapon, and others, provided additional avenues of commerce for the C&O Canal. These did not interfere with the path of canal, and were accessed by the river locks. On the Maryland side, streams entering the Potomac such as the Monacacy, Rock Creek, Seneca, Antietam, and Evitts Creek, required a culvert or an aqueduct. These feeder streams and rivers were, in the transportation parlance of the day, *laterals*. The same word was used by the railroads. As with the Potomac, some could be used for boat traffic, and some would need canalization.

We'll see how the canal interchanged with other canals at the terminus in Georgetown. As the C&O Canal became more real, other canals were planned to be built to intersect with it. This included one at the Monacacy in Frederick, and a canal to the Port of Baltimore. Baltimore was concerned that freight traffic from the west was being diverted to Virginia ports and ports in the District. The Merchants of Baltimore hedged their bets on supporting the canal and also the new untried technology of railroads. Baltimore has a good harbor, but lacks much in the way of navigatable rivers and streams, particularly from the west. This is what drove the Baltimore merchants in the direction of a railroad.

River Locks

Canal boats could enter the Potomac River at Georgetown via the Tide Lock. They would then need a steam tug to take them to their destination, there being no convenient place for the mules to walk. From 1876-1889, boats could later use the inclined plane a mile upstream of Georgetown to enter the river. This system used water-filled caissons to lower or raise the boat from canal to and from the river. This avoided the congestion along the Georgetown wharfs. From prior to the Civil War through World War I, coal boats were towed down the river to the Naval Coaling Station at Indian Head, MD. The Federal Government used 200,000 tons of coal from Cumberland in 1861, increasing to 1,000,000 tons in 1864. Most of this went to steam-powered warships, with a smaller amount used for manufacturing. The canal boat service to the Navy coaling station was critical in World Wat-I.

Boats could also enter the Potomac at Edwards Ferry, at Harper's Ferry, and at Shepherdstown.
Canal boats did not do well in the river, except in calm areas. At Edwards Ferry, boats carrying agricultural products from Loudoun County, Virginia, had access to the canal. At Harper's Ferry, the Shenandoah River trade was tapped. By means of the locks at Shepherdstown, the agricultural products of Jefferson County, Virginia, had access to the canal. Cement from the Boteler mill also went downstream. At Cumberland, boats could also enter the Potomac at its junction with Will's Creek. Any of the inlet locks behind the river dams allowed canal boats to pass to and from the Potomac as well. In the river, the boats could only maneuver by means of polling, or if they themselves were powered.

Of course, the Potomac River supplied the lifeblood of the canal which was the water. Sometimes too much, sometime not enough. At the downstream end, in Georgetown, local industry was supplied with water for power. This was tapped off the canal. Water usage was metered. For example, The Consolidation Coal Company installed water-powered hoisting machinery, similar to that used by the Swanton Coal Company at their facilities in Georgetown. The mills in Georgetown also used canal water as a power source. In 1837, Congress approved the sale of water power in the District by the Canal Company. The first use of this water for manufacturing was granted to George Bomford for his flour mill. Industries taking advantage of the power source included millers and textile manufacturers. The Company could not sell water power in Maryland fro milling grain, due to laws protecting the Jones Falls millers in Baltimore. This law persisted until the 1870's.

Water power rights were granted not only at Georgetown but at Weverton, Williamsport, and Hancock. At Seneca, water power was used to cut and dress locally quarried red sandstone. This became the source of the "brownstone" architecture of Georgetown, used in many of the warehouses for the canal trade. Casper Wever, a B&O Railroad civil engineer, planned a manufacturing town on 500 acres he purchased in 1834 along the Potomac downstream of Harper's Ferry. His model was the textile city of Lowell, Massachusetts. He purchased the land and the water rights, but a depressed economy delayed his plans. In 1849, a large cotton mill was built, along with a rifle factory and a marble works. When Wever died in 1849, the industries shut down. Water power was later used to run electrical generating plants at Dams 4 and 5. Water power was leased by the canal company; you could use the power, but you had to give the water back. It was gauged by the square inch. An iron plate with a square hole was used to meter the water. The rates were $2.50 per square inch per year, and no minimum quantity of water was guaranteed.

The Tenth Census of the United States, conducted in 1880, collected statistics on the use of water power. The report was issued in 1885. The report noted over 55,000 water wheels in use, producing over one and one quarter million horsepower, for a total of almost 36% of the total power used for manufacturing. The rest was animal or steam power. The use of steam power was increasing more rapidly than the others. Even though water power was "free" once the machinery was in place, steam power provided consistency. Water power, particularly from the Potomac was either too much or not enough. The greatest user of power was flour milling, followed by lumber, cotton, and paper manufacturing. The mills at Georgetown supplied with water by the Canal Company operated almost year-round.

The Patowmack Company

The Patowmack Company was created in 1785 to improve the navigability of the Potomac River. George Washington was the first president of the company. He served until he was elected President of the United States in 1789. Washington had realized the benefit to the nation of river navigation to the west, in his early surveys for Lord Fairfax of Virginia, and his presence on the Braddock Expedition against Fort Duquesne. He made at least six expeditions along the Potomac from 1754 to 1784. He owned land along the way and in Western Pennsylvania and the Ohio country. The Patowmack Company went on to build five canals around falls on the Potomac River. One set was at Great Falls, on the Virginia side. With these canals around the rough spots, the Potomac River could serve as a transportation route for most of the year. Washington also worried about the fate of the Trans-Appalachian lands. They were better connected to the North and South via the Ohio and Mississippi Rivers, than to the "United" States to the east. What was to keep them from forming their own country, and align with the Spanish, who then controlled the trade of Mississippi and the Port at New Orleans? Various schemes along these lines evolved.

Meeting at Washington's home at Mount Vernon in 1785, Representatives of Maryland and Virginia drew up the Mount Vernon Compact, establishing free trade on the Potomac.

The canal around Great Falls proved difficult to construct, but a success. Boats from Cumberland came down the Potomac with flour, whiskey, tobacco, and iron. Some boats were poled upstream with manufactured products such as nails, and seafood. The usual keelboat carried 20 tons of cargo. The downstream trip took 3-5 days; the upstream trip took 10-12. Most of the time, the boats were single-use, and sold for their wood in Georgetown. Most cargo was not worth the effort of polling it upstream. The westernmost part of Maryland had lots of trees. The boats entered the Potomac below Great Falls, and served the Port of Alexandria, just upstream form Washington's home at Mount Vernon.

The City of Alexandria was originally a 6,000 acre riverfront land grant from the Lt. Governor of the Virginia Colony to Robertson Howson in 1669. The land was transferred to John Alexander for six hogsheads of tobacco. A hogshead was a large wooden barrel, holding about 1,000 lbs. The laying out of the Town was authorized by the Virginia assembly in 1748, and the town was incorporated in 1779. George Washington was a member of the Board of Trustees

In the long run for the canal, expenses were excessive, and revenues were insufficient. Tolls couldn't cover the operating expense, let alone the construction debt service. Washington lobbied the States of Virginia and Maryland for the funds. James Rumsey, of steamboat fame, was appointed the Chief Engineer. In the end, it just didn't work. The Potomac Company surrendered its charter to the Chesapeake & Ohio Company in 1828. The Federal government was not providing funds for internal infrastructural improvements, and private financing was too expensive and hard to get.

One product from the era of the Potomac Company survived into the C&O Canal area. This was the surveys of the canal path west of Cumberland that were made in 1828, and validated in the 1870's. Discussion of extending the canal to Pittsburgh continued into the 20th century, using the original paths. The surveys were validated when railroads were built using those paths.

Alexandria Canal and the Aqueduct Bridge

The Alexandria Canal connected the City of Alexandria, Virginia, with the C&O Canal at Georgetown. Why, you would ask. Because the Port of Georgetown was silted up, and ocean-going ships, even coastal freighters, could no longer reach the wharfs. (This was a good thing in 1812, when the British Navy was prevented by shallow drafts and contrary winds from heading upriver to burn the cannon factory at Georgetown) The canal is on the Maryland side; isn't the Potomac River in the way? Yes, that's why they built an aqueduct across the Potomac. The aqueduct allowed for controlled access by canal boats. Mules can't swim and tow at the same time.

The Alexandria Canal Company was chartered in 1830. The Aqueduct Bridge was soon under construction. This would allow laden boats to bypass Georgetown, and proceed across the River to Rosslyn, and down the canal to the seaport at Alexandria. Construction of the system was completed by the early 1840's. The Alexandria Canal itself was seven miles long and had four locks, to accomplish a lift (or drop) of some 38 feet. The canal operated until 1886. The towpath bed became a trolley line, then present-day S. Eads Street. The Pennsylvania Railroad used a portion of the old canal route to construct a branch line to Rosslyn.

The Aqueduct Bridge was decked over during the Civil War, and heavily guarded. It was used for vehicular traffic, and had an extension built by the Great Falls and Old Dominion Railway (later, the Washington & Old Dominion Railway) for their trolleys. The bridge itself was closed in 1923 after construction of the parallel Key Bridge. The above-water section was removed by 1933, with bases of the pillars left in place to protect the Key Bridge from ice floes. Most of these were later removed as hazards to recreational boating on the Potomac.

Annapolis & Potomac Canal Company

This company was chartered in Maryland in 1828 to connect the C&O Canal with the Maryland State capital. The charter was amended in 1835, but nothing came of the effort. The State Capital did not want to be left out of the Canal phenomena, but at the same time didn't want to come up with the necessary funds for construction.

Chesapeake & Delaware Canal

The Chesapeake & Delaware Canal is located at the northern end of the Chesapeake Bay. It cuts across 14 miles of Maryland and Delaware to reach the Delaware River. From there, it is a short water journey south to the Atlantic, or up the river to Philadelphia. The canal project was surveyed in 1764, but construction did not begin until 1804. The canal was opened in 1829 using 4 locks.

Loaded C&O canal boats could be picked up in Georgetown by steam tugs, and taken downriver to the bay, then north to the C&D Canal. Once through that canal, the coal boats could proceed to Philadelphia, or south to the Atlantic, then up the coast to New York or New England. Demand for coal for industrial and heating purposes was great, and it was worth all this trouble at the time. The C&D Canal is still in operation, providing about 300 miles of short cut for ships to and from Baltimore. It handles about 40% of Baltimore's ship traffic, the rest coming up the Chesapeake Bay. It is administered by the Corps of Engineers. It has been widened and deepened to remove the need for locks, and is now a sea-level canal. The channel depth is now 35 feet and the channel width was increased to 450 feet, to support two-way traffic for most oceangoing ships. There is a railroad bridge crossing the canal, which is raised for canal traffic.

Frederick County Canal Company

In 1829, a branch of the C&O Canal up the Monacacy River near

Frederick was being considered. This would provide, with a connection between the Monacacy and the Susquehanna, an inland waterway path to New York. A survey was conducted in 1829 of the 24 mile Monacacy improvement project. But, the focus of Frederick quickly turned to the new railroad out of Baltimore, and the canal idea was shelved. The Monocacy is of limited use for boat traffic for transportation of goods.

The Monocacy River is the largest Maryland-side tributary of the Potomac. The Monocacy Aqueduct, which allows the C&O Canal to cross that river is the largest one on the canal system. The Monocacy rises near the Maryland-Pennsylvania border, and flows south. It now has an associated hiker-biker trail,

Maryland Cross-Cut Canal

In the early 1800's, when Baltimore was the Nation's largest exporter of grain, the merchants of the City looked for ways to expand their trade. The grain sources were located at the Ohio River. Piggybacking on the proposed construction of the Chesapeake & Ohio Canal, the merchants funded a series of studies, at least ten, for a way to connect the Port of Baltimore with the canal. These included canals branching off from the C&O and extending across Montgomery and Howard County, or using the Eastern Branch (Anacostia River), which is, unfortunately only about 8 miles long, and does not reach Baltimore. A reliable source of water for the canal was a problem. The lack of a definitive canal solution, and the cost, drove the merchants of the City of Baltimore to back a newer technology solution: the railroad.

In 1836, the Maryland Assembly supported several internal improvements. $3 Million went to the B&O Railroad and to the C&O Canal. Additional money went to the Baltimore & Susquehanna Railroad, the Annapolis and Elkridge Railroad, and the Eastern Shore railroad line. More importantly, $500,000 went to the newly formed Maryland Canal Company, to provide surveys for the access to Baltimore from the C&O Canal. Nothing ever came of the efforts.

Pennsylvania Canal

The canal and railroad system in Pennsylvania, the Pennsylvania *Main Line of Public Works*, was a 19th century transportation system funded by the Commonwealth of Pennsylvania. Both railroads and canal were used, as well as hybrid systems, such as the Allegheny Portage railroad. The Main Line ran from Philadelphia to Pittsburgh, connecting the eastern seaboard with the Ohio River. Construction on the system began in 1826, and was completed by 1834. It was sold to the Pennsylvania Railroad Company in 1857.

The Allegany Portage Railroad allowed for continuous boat traffic over the Allegany Mountains. It was 36 miles long with eleven inclined planes, equipped with stationary steam engines. The highest summit was 2,322 feet above sea level. The railroad hauled canal boats up over the mountain, and let them down the other side. It was a complex and slow operation, but cheaper and faster than tunneling or a using a series of locks in a staircase fashion.

The Western Division Canal, from Johnstown to Pittsburgh, was the destination of the Chesapeake and Ohio Canal. At Pittsburgh, the Western Division Canal had a turning basin, and the Allegheny Outlet provided a connection into the City. An aqueduct more than a thousand feet long got the canal across the Allegheny River. The canal also went through an 800-foot tunnel to link with the Monongahela River. The first boat from Johnstown to Pittsburgh was in 1831. The canal was 104 miles long, with 68 locks, and 16 aqueducts. Canal boats from Georgetown were intended to be able to reach Pittsburgh and the Ohio River in the original plans of the C&O Canal.

Potowmack Canal

The Potowmack Canal was a skirting canal for the Great Falls of the Potomac, built by the Potowmack Company. It took 17 years to build, being completed in 1802. It stalled for a while for lack of money, solved by an infusion of cash by the State of Maryland in 1799. Five locks were required to negotiate the drop in elevation at Great Falls. Extensive black powder blasting was required to form a path through the rock cliffs. Before the locks were completed, an inclined plane and winch were used to lower cargo, such as barrels of flour. Just the canals around Little Falls and Great Falls were a vast improvement to Potomac navigation.

Shenandoah Canal

The Shenandoah Canal at Harper's Ferry, WV, was excavated by the Potowmack Company north of Virginius Island in 1807. It was only 580 yards long, but provided water access to the industries on the north end of the island, where a pulp mill was built.

Virginius Island was an Industrial enclave in the Shenandoah River, just south of Harper's Ferry. The island contained an elaborate series of tunnels to supply water power to turbines. Water was drawn from the Shenandoah through a series of control gates into a holding basin for distribution to the various industries. In 1848, a large cotton mill was built; followed by a second one in 1849. The Shenandoah Pulp Mill, built in 1888, was driven by ten turbines, and had a capacity of 40 tons per day. The Island also supported Hall's Rifle works, circa 1820, a tannery, a cooper shop, and an iron foundry. Workers' housing was also located at the facility. The island facility produced material and supplies for the nearby Federal Armory. The Harper's Ferry and Shenandoah Manufacturing Company operated at the facility, circa 1850. Virginius Island is part of the Harper's Ferry National Historical Park.

Washington City Canal

Included in the proposed plans for a new capital city was L'Enfants' design for an elaborate city canal system. George Washington was impressed by his designs. Temperamental artiste L'Enfant later stormed off the project, taking his designs with him, but the work was continued and completed by Andrew Ellicott. The canal was seen as an excellent avenue of commerce for the Capital City, particularly for the fur and pelt trade. The new Capital of the United States was viewed not only as the seat of Government, but also as a commercial center.

Different parts of the City viewed the canal differently at the time. The Citizens of Georgetown were pleased to be the port for the planned Chesapeake & Ohio Canal. Merchants in other parts of the City needed a way to minimize local delivery charges.

The canal may have been marked out by Washington himself. L'Enfant realized the importance of a City canal system. It would be bordered by wide boulevards on each side. In 1802, by Act of Congress, the Washington Canal Company was chartered. The charter was signed by President Jefferson. Financing remained a problem, and construction was delayed by the invading British Army in the War of 1812. The canal finally opened in 1815. The canal connected with the C&O canal at Georgetown in 1833. Canal boats from upstream made their way along what is now Constitution Avenue towards the Anacostia. The canal carried marble from a quarry near Monocacy directly to the building site of the Capitol building.

In 1831, the City of Washington purchased the Canal from its private investors, for $50,000. Improvements and maintenance were expensive. The C&O Company was coerced to complete the connection between the canals by the withholding of funds from the City. The C&O connection started at the Rock Creek Basin, and followed 27th Street to Constitution Avenue to Seventeenth Street, and the connection with the Washington Canal. By the

1850's, 120 tons of Cumberland coal per boat went through the City, past the not-yet-completed Capitol. The boats would pass on to the Navy Yard, and general merchandise would be loaded at the Center Market.

By the late 1870's the City Canal had become little more than an open sewer, and the section where Constitution Avenue runs today was filled in. The section at what is now Canal Street was filled in some years later. At Seventeenth Street and Constitution Avenue, a stone lock house from the City Canal still stands, in the shadow of the Washington Monument, and across from the White House.

The Potomac River

Technically, the Potomac is formed by the confluence of the North Branch and the South Branch at Green Spring, WV. It then flows to the Chesapeake Bay tidewater below Washington, DC. In the western part of Maryland and West Virginia, it is common to refer to the North Branch just as "the Potomac."

The South Branch of the Potomac River

The South Branch of the Potomac Rivers flows northward to combine with the North branch near Green Spring, WV. The South Branch is paralleled by the current South Branch Valley Railroad. The most scenic part of the River is near the northern end, where it squeezes between mountains some 2,000 feet high to the west, and 1,600 feet to the east. This section is a protected American eagle habitat. Passing through this area on rafts down the river, or on the scenic *Potomac Eagle* tourist train, nesting and flying eagles are almost always seen.

The North Branch of the Potomac

The North, or main Branch of the Potomac rises in the mountains of West Virginia, at the western border with Maryland. The point is marked by the Fairfax Stone. The boundary between the British

Colonies of Maryland and Virginia, set by Royal Charter, was to be the Potomac River, but it was not clear which branch of the river. This was resolved in 1785 to be the North Branch. The State of Maryland "owns" the river, to the low water mark on the Virginia shore, but Virginia has rights of use to the water. The details have been in dispute for more than 400 years, illustrating the problems with granting rights by Royal decree concerning territories an ocean away, and poorly mapped. Recent disputes have gone to the U.S. Supreme Court.

Will's Creek

Will's Creek originates in the Allegheny Mountains of Pennsylvania, flows south, and joins with the Potomac River at Cumberland. It passes through the scenic Narrows before flowing into Cumberland. Due to repeated flooding of the city, the US Army Corps of Engineers channelized Will's Creek from the south side to the Narrows, through Cumberland, to its confluence with the Potomac in the 1950's. The path of Will's Creek was considered for the northern route of the C&O Canal path from Cumberland to Pittsburgh. This would have crammed the canal into the Narrows, which already contained Will's Creek, two railroad lines, and the National Road.

The Ohio River

The Ohio River is formed by the confluence of the Allegheny and Monongahela Rivers at the current location of Pittsburgh, PA. From there it flows close to a thousand miles to join with the Mississippi at Cairo, Illinois. Why the river from there to the Gulf of Mexico at New Orleans uses the name of the Mississippi and not the Ohio, when the latter is the larger at that point, is unclear.

The Ohio and its upstream tributaries were a major transportation route for the Native Americans, and the river system was explored by the French in the later 1600's. With some short portages, a canoe could travel from the St. Lawrence at Quebec to New

Orleans, both French territories at the time. This provided a major north-south transportation pathway for trade and empire. France built a series of fortifications along the Ohio and up to Quebec, to put notice to the British Empire that the territory was already claimed and would be defended. The British, moving westward from the seacoast, disputed this. Portions of a World War, the 7-Year's War, between these Empires were fought for access to the Ohio.

What's the big deal? Why were the railroads, the Federally funded National Road, and the C&O Canal trying to get to the Ohio River? After the Revolutionary War, the United States' western boundaries were somewhat ill-defined. The territories west of the Appalachians could become new states, new countries, the property of the Spanish Empire, or maybe even revert to the Native Americans. The territory did become part of the fledging United States, and was rich, fertile farmland. The Ohio River provided cheap and available transportation north-south, but a connection to eastern tidewater was needed. This was recognized by George Washington, even before the 7 Years War, what is called the French and Indian War in the United States (they both lost). Today, using satellite imagery, the path is obvious. Cumberland, MD, represents a point along the north-south mountain range where an east-west transportation route could be accomplished, to link the tidewater with the Ohio. Money. Trade. Technology. It comes together there.

Railroads

This section gives a brief background on the railroads that touched and interacted with the C&O Canal, listed alphabetically. For further information, see the references at the end of this document. Don't bother reading this section end to end, but refer back to it as the relations between the railroads and the canal are discussed in later chapters.

Some of these railroads discussed are still running, and some have been out of business for a hundred years. Some of these railroads interchanged cargo with the canal. Some merely crossed it on bridges, or ran along it. Most were standard gauge (spacing between rails of 4 feet, 8 ½ inches); some were narrow gauge. Canal passengers arriving on packets, or by casual transportation on freight boats, could use the extensive DC trolley system to get around the City. Originally pulled by horses, the trolley system evolved to electric motive power, although an air-powered system was tried for a while. The multitude of DC trolley were consolidated, and mostly replaced by diesel buses. Later, the DC Metro system, mostly running underground, augmented the buses. The Metro does not serve Georgetown.

There are several railroad lines near the canal you can ride today, including Amtrak from Union Station in DC through Cumberland to Pittsburgh, the Western Maryland Scenic Railroad in Cumberland, and the Potomac Eagle Excursion on the South Branch Valley at Romney, WV. In addition, the MARC Commuter rail line uses CSX track from Martinsburg, WV to Union Station in Washington, D. C., traveling along the Canal for mot of the way.

There is a classic picture by artist H. D. Stitt of a locomotive on the B&O line spooking a horse drawing a canal boat. There was animosity between the railroad and the canal, but cooperation as well. The fate and fortunes of the railroads and the canal were intertwined in a byzantine manner.

Amtrak

Founded in 1971, the National Passenger Rail Corporation operates rail passenger in the United States, and specifically the Capitol Limited from Washington, DC through Cumberland once per day in each direction. This line generally follows B&O line along the Potomac River on the northern bank, or Maryland side, until Harper's Ferry. There, the rail line emerges from a tunnel to cross the canal and the Potomac on a bridge. It swings south to Martinsburg, WV, and stays south of the river until crossing back into Maryland before Cumberland. From Evitts Creek east of Cumberland, the Canal path is picked up again. The route from Cumberland to Pittsburgh generally follows the northern proposed path for the canal extension to the Ohio River.

Baltimore & Ohio

The B&O Railroad and the C&O Canal were in a race, starting on July 4, 1828, to the Ohio River. This race was, in its day, as significant as the space race in the 1960's. The railroad won the race. However, there was more cooperation than competition between the Canal and the B&O Railroad when it came to shipping product. There was more than enough coal to go around. This culminated in the B&O buying and operating the canal in 1889. Not that the rail company wanted a canal; but they certainly did not want one of their rivals to have it.

The Baltimore & Ohio Railroad was perceived as the C&O Canal's chief competitor. The Canal never recovered from the B&O's reaching Cumberland first. The two massive construction projects vied for the same pool of unskilled and skilled laborers, and for the same construction funds.

At the east end of the canal, the B & O Railroad served Washington DC from the West after 1873. At the far west end, the B&O went on to its goal of the Ohio River. On the way, it went through Piedmont, Virginia (later, West Virginia), opposite the confluence of the Potomac and George's Creek. The George's Creek Region was the major source of coal in Western Maryland.

From Point of Rocks, the B&O line reached DC after the Civil War. Before then, DC traffic had to go up to "the old main line" at Relay, south of Baltimore. The point of controversy was at Point of Rocks, a choke point east of Harper's Ferry, and the canal reached there first. Long, legal battles ensued. Finally, the B&O, in exchange for buying 2500 shares of Canal stock, got the right to build to Harper's Ferry. This was the "Compromise of 1833." But the railroad rapidly built south and then west, and reached Cumberland first.

In 1910, the B&O completed the Georgetown Branch, a freight spur from the Metropolitan Branch at Silver Spring, MD. This provided weekly coal service to Georgetown. Service continued until 1985. The right-of-way for this line now serves as the Capital Crescent Trail, a Rails-to-Trails project.

The B&O had higher aspirations for the Georgetown Branch than weekly freight runs. It was supposed to be a link to the southern railroads, crossing the Potomac at Georgetown, west of Chain Bridge. The existing rail link, the Long Bridge, was owned by the B&O's rival, the Pennsylvania Railroad. The B&O did have its Alexandria Branch line, but the Penny refused to interchange freight with the B&O in Virginia. Besides, the Alexandria Branch relied on a ferry system, not a bridge. The Pennsy also opposed the construction of a bridge by the B&O.

Later, the development of the Potomac Yard gave the B&O the access to Virginia that it wanted. The line to Georgetown was completed, including a bridge over the canal and Canal Road near Arizona Avenue. In Georgetown, the rail line provided coal to the power house of the Capital Traction Company (circa 1912). The line also has the distinction of carrying limestone for the construction of the Lincoln Memorial. When the Capital Traction generating plant closed in 1933, the only coal customer of significance in Georgetown was a Government Services Agency heating plant. This plant switched to coal delivery by truck in 1985.

The B&O went into receivership in the depression of 1893 for 3 years. It was then controlled by its arch-rival the Pennsylvania Railroad for 10 years. The Pennsy had fought hard to keep the B&O out of Pennsylvania, and was foiled only by the B&O's purchase of the Pittsburgh and Connellsville Railroad by the B&O.

One more brief highlight for the Georgetown Branch - it saw the running of the Smithsonian's replica of the pioneering locomotive *John Bull* on the locomotive's 150th anniversary.

Billmeyer Lumber Company Railroad

The Western Maryland Railway reached Little Orleans in 1904, building on the North side of the Potomac. The Billmeyer Lumber Company built a railroad siding at Little Orleans along the Western Maryland line. The lumber traveled from the company's mill to the railroad siding by horse and wagon. At Little Orleans the lumber was loaded onto rail cars on the siding. Depression in the lumber industry forced the bankruptcy of the Billmeyer Lumber Company in 1927.

The WM had two side tracks in the town, one 553 feet long, and the other 752 feet long. The second siding, formerly belonging to Billmeyer, was purchased by a local resident and was later used to load coal into hopper cars. This side track still exists.

Bloomington and Fairfax Railroad

This line was chartered in 1880 by Senator Elkins of West Virginia to extend from Bloomington to Elk Garden, WV at the western end of the Potomac. The North Branch of the Potomac rises from a spring at the Fairfax Stone, high in the mountains of Maryland at the West Virginia border. The rail line was to be 8 miles long, and probably narrow gauge. It was intended to haul coal from mines in the area. There is no evidence that it was ever built. If the canal extension had been built on the southern route, this line would have been of importance as a coal feeder.

Chesapeake & Ohio Railroad (Chessie System)

In 1962, the Chesapeake & Ohio (C&O) Railroad acquired stock ownership of the B&O, leading to the creation of the Chessie System as a holding company. The B&O completed full control of the Western Maryland Railway in the 1970s. The C&O and the B&O unified the operations of the WM into the Chessie System on January 15, 1973. The operations of the various lines were consolidated, and engines were repainted into a new bright corporate look, but with the initials of the predecessor roads in small letters painted under the windows. Chessie engines hauled the coal trains along the old B&O and WM lines. Parallel trackage was abandoned in favor of the better of two routes. The Western Maryland line at Cherry Run, near Hancock, was deemed redundant, and abandoned. It is now the Western Maryland Rail Trail, for hiking and biking, paralleling the C&O Canal towpath in the area. Chessie merged with the Seaboard Coast Line (SCL) to form CSX Corporation in 1980. *Chessie was* the name of a cute sleeping kitten featured as the logo in the railroad's ads for passenger service.

Conrail

Formed in 1976 from the bankrupt Penn Central Railroad, the consolidation of the New York Central System and the Pennsylvania Railroad, Conrail acquired the lines and equipment of those predecessors. Conrail was jointly acquired by Norfolk Southern and CSX Corporation in 1998. Several lines had been built across Maryland (and the canal) from North to South by the Pennsy. These then became Conrail lines.

CSX

Present-day CSX Corporation is a class-1 railroad, one of four in the United States. The CSX Corporation is the surviving entity of all of the Western Maryland shortlines, as well as the B&O and the Chessie system. Today, multiple 200 ton diesel electric units of CSX can be seen lugging unit trains of coal more than a mile in length along the original rights-of-way.

In 1980, Chessie and the Seaboard System merged to form CSX Corporation. In 1986, CSX Transportation (CSXT), a division of CSX Corporation, was formed. The Cumberland Coal Business Unit was set up by CSXT in 1993, and headquartered in Cumberland. This business unit was responsible for the shipment of over 18 million tons of coal in 1996. Over 850 route-miles of track from West Virginia mines to the port at Curtis Bay, Baltimore, are covered. All of the coal passes through Cumberland, as do the returning empty hoppers. The engine and car service facility in South Cumberland was originally built by the Baltimore & Ohio railroad in 1892 as a steam engine servicing and repair facility. Now, it is used to service and repair diesel locomotives, to provide training, and to service and classify freight cars. The volume of repair work is staggering. The car shops see over 2,200 freight cars per month. The diesel shops see 800 units of CSX's motive power fleet per month. The shops and adjacent facilities employ 900. The shop services diesel locomotives from EMD Corporation and General Electric.

There is a Baltimore Service Lane freight classification yard, with a 32-track hump facility along the canal. The east throat of the yards is marked by Mexico tower. The west end of the yards feed into the Baltimore street crossing in Cumberland. The yards are busy places day and night, 365 days a year. On a given day, 50-60 trains pass through Cumberland. Long cuts of empty coal hoppers are assembled to go west. Loaded auto racks go east. Lumber and coke come in from West Virginia. Industrial chemicals and lumber cars get added. On a given day, one can see examples of just about any of the many types of freight that can be hauled on a railroad go through Cumberland. The eastbound receiving yard stretches along the canal from the old Pittsburgh Plate Glass factory to Evitts Creek.

CSX's National Gateway Project is a public/private partnership to increase the capacity of the rail lines. Some of this involves adding more track. Other areas need the track, bridges, or tunnels modified to allow double-stack containers in Intermodal traffic to flow more efficiently.

Cumberland & Pennsylvania Railroad

The Cumberland & Pennsylvania Railroad operated in Allegany County, MD, except for short forays on Pennsylvania Railroad track in PA, and on B&O track in West Virginia. The C&P was a railroad of the age of steam. A single gasoline-electric car was used quite late in the railroad's operating life for mail and passenger service. The C&P built and maintained its own equipment at its facility in Mount Savage. The railroad was chartered in 1850, and for most of its life, was owned by the Consolidation Coal Company. The C&P was acquired by the Western Maryland Railway in 1953. The author's grandfather retired from the C&P just before the acquisition.

The C&P interchanged with the B&O at Cumberland and Piedmont, crossed the Western Maryland at Westernport, and met the Pennsylvania Railroad at State Line, just north of Ellerslie, MD. It was also a major supplier of coal to the canal. The major facilities were in Mt. Savage, the heart of the railroad. The office building, built in 1902 of enameled brick, still stands. The brick round house, circa 1907, had a sixty foot Armstrong turntable. The stone machine shop and car shop from 1866 survive.

The C&P provided the mail and the railway express service to Frostburg and the mining communities of the George's Creek. Twelve passenger stations were located along the line, with a shared station at Cumberland. The only surviving example is the station at Frostburg, which survived as a specialty restaurant (now closed) for the Western Maryland Scenic Railroad. The station in Piedmont, shared with the B&O, partially survived as the first floor of a formally two story structure.

There was an 1871 agreement between the C&P and Pennsylvania's Bedford and Bridgeport Railroad which allowed Cumberland coal to be shipped to Philadelphia. By 1872, Cumberland coal was going to the seaport of South Amboy, NJ, in direct competition to the Baltimore & Ohio. The C&P also hauled coal to the C&O Canal boat loading facilities at the basin in Cumberland.

The C&P provided the region with a home-grown transportation infrastructure; it enabled people in the outlying communities to go to market, and to attend school in the cities. Passenger service was provided, and made connection with the B&O at Cumberland and Piedmont. Combination tickets were popular. These provided round trip transportation and a ticket to the popular Academy of Music in Cumberland. Special trains on Sundays provided transportation to Church and social events. Baseball games and Fourth of July Celebrations also were served by special trains. Narrows Park in LaVale was a popular weekend picnic destination.

The Main Line of the C&P extended 31.8 miles from Cumberland to Piedmont, via Frostburg. There was one tunnel, under the town of Frostburg, still extant. The Eckhart Branch Railroad, built as the Maryland Mining Company's Railroad, was acquired in 1870, and operated into the middle 1950's. The State Line Branch headed north from Mt. Savage Junction to meet the Pennsylvania Railroad at the State Line.

Cumberland & Piedmont Railroad

(See, Piedmont and Cumberland Railroad) Railroad companies of the time were frequently changing names and incorporating in different States, as economic and political situations changed.

Cumberland Valley Railroad (MP 97.3 on canal)

This little line (not located near Cumberland, MD) didn't cross the canal or the Potomac until after the Civil War.

In 1873, the CVR extended south from Hagerstown, MD, to the Canal and the Potomac River. In 1874, the line was completed from Harrisburg, Pennsylvania to Winchester, Virginia. It crossed the canal and the Potomac below Williamsport. The C&O Canal sought to facilitate the operation of this line, to expand the economic benefits of both the canal and feeder railroad. The Canal constructed wharf facilities for coal, lumber, and agricultural products.

During the Civil War the rail line had strategic importance in supplying Union troops in the Shenandoah Valley. It also ran the first passenger sleeping car service in the U.S. Service was provided to Philadelphia and Pittsburgh via the Pennsylvania Railroad. In 1889, it reached Martinsburg, WV, and Winchester, VA, providing access to the Shenandoah Valley.

In June, 1882, the Shenandoah Valley Railroad was opened from Hagerstown to Roanoak Virginia. In conjunction with the Norfolk & Western Railway the Cumberland Valley Railroad operated the middle link of the New York-Harrisburg-Hagerstown-Roanoke, Virginia, passenger trains.

The Pennsylvania Railroad gained control of the Cumberland Valley line around 1859, and officially purchased it on June 2, 1919. Loading of coal from canal boats to railroad cars was done by crane until 1924, for shipment to Pennsylvania and Virginia. The Pennsy's successor, the Penn Central, closed all railway facilities in Chambersburg in 1972 and its successor, Conrail, abandoned major pieces of the line in 1976.

Eckhart Rail Road

The Eckhart Rail Road went from the Eckhart mines east of Frostburg in Western Maryland, to the B&O railroad at Cumberland. Later, their Potomac Wharf Branch allowed access to riverboats and canal boats. The Eckhart Rail Road eventually stretched from the B&O's Queen City Station in Cumberland 12.4 miles west to the mine head at Eckhart. This was the location of Consolidation Coal Company Mines 3, 4, and 10.

The rail section from Eckhart to Will's Creek was completed in 1846 by the Maryland Mining Company. The Cumberland Coal & Iron Company, chartered in 1850, purchased the Maryland Mining Company's mines and railroad property (including the village of Eckhart) in April 1852. The rail line was extended to the nearby Hoffman mines in 1859. Cumberland Coal & Iron was in turn acquired by the Consolidation Coal Company in 1870.

The Eckhart Rail Road was used to carry coal to flat bottom Potomac River boats and to canal boats, before the canal wharf facility was completed. For a period of 20 years, from 1850-1870, the Eckhart Rail Road operated independently. It then became the Eckhart Branch of the Cumberland & Pennsylvania Railroad.

Frederick and Pennsylvania Railroad

Ground was broken for this line in 1869 near Woodsboro, MD. It reached Walkersville and then Frederick, bringing a means of transportation to the area, and increased prosperity. Later, it would become part of the Pennsylvania Railroad Frederick Secondary line. Now, a surviving portion hosts the Walkersville Southern Tourist line. Had the earlier lateral Monocacy Canal been built, it would have served the same area.

George's Creek & Cumberland

The George's Creek and Cumberland Railroad was born out of the competition that smaller coal companies had with Consolidation Coal's captive hauler, the Cumberland & Pennsylvania Railroad.

The George's Creek & Cumberland Railroad was the child of two mining companies, the Maryland Coal Company, and the American Coal Company. In the George's Creek Coal Region of Allegany County in the 1870's, the transportation monopoly was controlled by the Cumberland & Pennsylvania (C&P) Railroad, which was owned by the Consolidation Coal Company. Rival companies could not get competitive rates to move their coal from the mines to the B&O and the canal. The solution was two-fold: build a second railroad, and involve the B&O's rival, the Pennsylvania Railroad. The GC&C should not be confused with the earlier George's Creek Rail Road, built in 1853 to allow shipment of iron products from the furnace at Lonaconing to the planned canal and railroad terminus at Westernport That line became part of the C&P.

The GC&C was born out of controversy and competition with the C&P, and this climate of antagonism continued. The GC&C had to fight its way past the C&P into Cumberland, and then fight for the right to reach the canal over B&O trackage. The first fight was at the west end of Cumberland, an area known as City Junction. The GC&C had to cross the C&P's Potomac Wharf Branch, which was there first. The Pennsylvania Railroad in Maryland line had been built from the Pennsylvania state line to the west side of the Narrows. It was their intent to continue down the north side of the Narrows, along with the C&P and B&O mains, to Cumberland. The C&P persuaded the other road to bridge Will's Creek, and continue down the south side of the Narrows, then cross the C&P's Potomac Wharf Branch at City Junction. When the line was built to City Junction, the C&P conveniently changed its mind. It kept an engine parked at the intended crossing point, blocking construction. When the engine was a bit late getting back into position one day, the GC&C trackmen forced the crossing. The C&P trackmen tore it out. Tempers flared. When the C&P raised its trackbed, it made the crossing impossible. The final issues were decided in court, in favor of allowing the GC&C crossing with proper compensation to the C&P. Then, the B&O did not want to grant the GC&C trackage rights to reach the canal terminus, and this issue also had to be resolved in court. The GC&C was controlled by the American Coal Company and the Maryland Coal Company, rivals to the Consolidation Coal Company which had locked up most of the smaller mines, and owned the C&P Railroad.

The Western Maryland Railroad purchased the controlling stock interests of the GC&C on January 17, 1907. The GC&C was a small but critical part of the Gould master plan for a transcontinental railroad link. The financial panic of 1907 put an end to these grand schemes. Bankruptcy followed. Operation of the GC&C was taken over by the newly reorganized Western Maryland Railway in July of 1913. A full merger and consolidation took place on January 23, 1917. The line was operated until 1939, when the Western Maryland abandoned the track from George's Creek Junction to Midland. Mines west of Midland were then served through the interchange with the old C&P line at Jackson Junction, north of Lonaconing.

From 1869 to 1879, James A. Millholland, son of the James who set up the C&P shops, was the second vice-president of the C&P. He was lured away to become General Manager, later President, of the George's Creek and Cumberland Railroad. Part of the deal was his new house, located behind the Emmanuel Episcopal Church on Washington Street in Cumberland. This spectacular Victorian structure is still standing, and is used by the church.

The GC&C started as two separate pieces: the line to Vale Summit and Lonaconing called the GC&C, and the connection to Pennsylvania, called the Pennsylvania Railroad in Maryland. The two pieces were merged under the name George's Creek and Cumberland. Later, the Connellsville Extension of the Western

Maryland Railroad was built under the umbrella of the GC&C. On July 1, 1913, the GC&C was formally absorbed into the Western Maryland System, and the Connellsville extension became WM trackage, as did the portion called the Pennsylvania Railroad in Maryland.

George's Creek Rail Road

The George's Creek Coal & Iron Company was formed in 1835, and chartered in the State of Maryland on March 29, 1836. Between 1837 and 1839, the company built an iron furnace at Lonaconing. The furnace, fueled by coke, went into blast in 1839. There was plenty of iron ore, limestone, water, and coal locally, but the major problem the company faced was transporting finished products to market. Production reached 75 tons per week, and local iron needs were quickly satisfied. Some products were shipped out by wagon, including such items as dowels for the C&O Canal walls. The adjacent casting house made farming implements, mine car wheels and track, and household utensils. The furnace output was in the form of pig iron, which was sold to be recast, or worked by foundries or blacksmiths.

With production going well, iron piled up in Lonaconing. In 1842, sales of pig iron to foundries in Cumberland were begun, with delivery by wagon. An adjacent sawmill and lumberyard, also owned by the company, recorded sales to the Mt. Savage Iron Works, then involved in building their own furnaces. In the fall of 1842, pig iron was offered to the B&O railroad at a price of $29 per ton. Delivery was still a problem. After experimenting with a horse powered tram road, the company realized that a rail line, built down the George's Creek Valley for 9.2 miles to the Potomac River at Westernport, would be the answer to the transportation issue. The rail line was finished from Lonaconing to Piedmont in 1853, where it connected with the recently arrived B&O Railroad. It was, unfortunately, too late to provide the needed market access for the Lonaconing Iron Furnace. The furnace in Lonaconing was abandoned in 1855. Coal, not iron, became the most important commodity shipped out of the region. The furnace produced 1,860 tons of pig iron in its last active year. Harvey states that the furnace facility was too technologically advanced for its time. However, it provided an impetus for the mining industry and for the construction of the railroad, and served as a model for a similar iron working facility built at Mount Savage.

The furnace complex was visited by the Superintendent of Construction for the B&O, Casper Wever, in June of 1839. Shortly afterward, the shareholders of the C&O Canal visited. With the furnace up and operating, the facility expansion plans included a forge and rolling mill. The facility supplied strap iron and dowels to the Canal. The furnace facility began to concentrate on the railroad to meet with the canal or the railroad at Westernport. In 1850, surveys were complete. The B&O reached Piedmont, across the Potomac River from Westernport, in July of 1851. In September of that year, the railroad construction began up the George's Creek. The railroad was opened on May 9, 1853. In June, a total of 1,061 tons of coal were shipped. In all of 1855, 225,000 tons of coal were shipped, sometimes in 102 car trains. Today's unit coal trains are 100 or 120 cars. Iron, ore or cast, did not figure into the shipments. In 1856, the line was extended from Lonaconing northward to connect with the C&P from Frostburg. The George's Creek Coal & Iron Company's 9.2 mile railroad was acquired by the C&P on October 23, 1863. The shops and engine house at Lonaconing were used until 1867. In 1991, the George's Creek subdivision of CSX hauled 195,197 tons of coal over this line, as compared with 225,000 tons in 1855. The line was abandoned by CSX, and recently acquired by the new George's Creek Railway, for the purpose of reopening the coal trade.

Had the canal extension been built at least as far west as Westernport, the George's Creek would have been a main artery for coal from the George's Creek Valley to the canal.

Georgetown Barge, Dock, Elevator, & Railway Company

This entity of the Baltimore & Ohio built a mile of track in 1889 along the Georgetown waterfront from Rock Creek to the Aqueduct Bridge. It connected to the B&O's Georgetown Branch in 1910.

Green Ridge Railroad

The Green Ridge Railroad was a 3-foot guage line in the eastern part of Allegany County, running roughly north-south. Two engines were built in Mt. Savage, MD, under contract to T. H. Paul of Frostburg, for sale to Mertens, the owner of the railroad.

The Green Ridge Railroad was built by the Mertens Company in 1889. It initially consisted of 16 miles of track. The railroad made connection with the B&O Railroad at Okonoko, WV, using a trestle over the canal and the rail line, and with the C&O Canal at Darkey's Lock (lock 67). By 1896, the line had grown to 26 miles of track, and had a passenger car. The rail line was closed in 1897.

Green Ridge locomotive Number 1 was featured in the Mt. Savage Works Catalog, as the model for the 0-6-0 units. GRRR Number 2 was a 0-4-0 unit. The rail line was eight miles to the east of Cumberland, in the vicinity of Town Hill, and Fifteen Mile Creek. It hauled lumber to a sawmill at Oldtown. The lumber was used by the Merten's boatyards in Cumberland to construct canal boats. The railroad operated from 1889 to 1897.

Kulp Lumber Company Railroad

The Kulp Lumber Company, incorporated in 1895, operated a sawmill in Oldtown, Maryland. The rail line, used to bring logs to the mill, was narrow gauge. The locomotives were Climax type geared locomotives, favored for logging operations. The railroad was also used to deliver finished product, sawn lumber, to a connection with the Western Maryland Railway, which is on the north side of the Potomac and canal at that point. The operation was fairly extensive, with company housing, a round house and locomotive servicing facility, and the steam-driven circular sawmill, with a capacity of 50,000 feet of lumber per day. The company was a major source of employment for Oldtown. The rail line approached lower Town Creek Road, and then went upstream along Town Creek to Flintstone. There were numerous spurs off the main line servicing cutting locations. The Company never recovered from the death of founder Monroe Kulp in 1911, and closed in 1914.

MARC

The Maryland Rail Commuter service (MARC) is an integral component of Maryland's transportation system. The 187-mile system serves as a major means of commuting between Washington, D.C. and Baltimore and Perryville, Maryland as well as Washington, D.C. and Martinsburg, West Virginia.

MARC's Brunswick line runs Monday through Friday from Union Station in Washington, DC as far as Martinsburg, WV. This line is shared with CSXT and Amtrak. The maintenance yards for MARC diesel locomotives are at Brunswick, MD, adjacent to the canal. Brunswick is an ex-B&O facility.

MARC began operating in 1974. MARC trains are operated by Amtrak personnel on the electrified Penn Line and CSX Transportation on the Camden and Brunswick Lines under contract to the Mass Transit Administration of the Maryland Department of Transportation. The MTA acquired control of MARC under legislation by the Maryland General Assembly in 1992. MARC had been providing service throughout the Baltimore-Washington metropolitan area for 18 years at that time.

The State of Maryland began subsidizing commuter rail service in 1974 on what are now the Camden and Brunswick lines through a contract with the B & O Railroad (now CSX). In 1976, the state began subsidizing service on what is now the Penn Line with Conrail. The State Railroad Administration - a separate entity under the Maryland Department of Transportation - was created by executive order in 1978 to oversee these commuter rail contracts as well as freight rail operations in the state. Amtrak took over operation on the Penn Line in 1982. The name MARC and the decision to begin marketing the commuter rail system began in 1984.

The Brunswick follows the old B&O Metropolitan (Met) Branch and the canal from Washington to Point of Rocks. Then, it follows the old B&O main stem to Martinsburg, WV. Construction on the Met line started in 1866 east from Barnesville, MD. In 1928, it was double tracked. In that year, 38 passenger trains per day traveled the Met, plus heavy freight traffic. Today, typically this line daily sees between 15 and 25 CSXT freight trains, 18 MARC trains, and Amtrak's Capital Limited. This line has 3 station stops in West Virginia, 13 in Maryland, and a destination of Union Station in Washington DC. At Union Station, MARC comes together with Amtrak, the Virginia Railway Express, and the D. C. Metro System.

Maryland Mining Company Railroad

The Maryland Mining Company (MMC) was incorporated in Maryland on March 12, 1829. The company built the railroad from Eckhart to Will's Creek at Cumberland, a length of 9 miles, and later extended the line as the Potomac Wharf Branch. The original line was built by a company in which Henry R. Hazelhurst was a partner. Originally, he worked for the B&O Surveyor, Knight, drawing maps. He was a cousin and brother-in-law of Benjamin Latrobe. Hazlehurst surveyed several of the B&O lines. Later the B&O Corps of Engineers was "under the direction of H. R. Hazlehurst, Esq." The Cumberland *Alleganian* newspaper of August 30, 1845, reported that the Maryland Mining Company Railroad "was constructed by Gonder & Hazlehurst Co. who completed 9 ½ miles in 3 months." It was ¾ complete, with 2 tunnels remaining. The total rise was 1117 feet from Cumberland to Eckhart. Five miles were at 130 feet per miles (2.5%)." In contrast, the B&O's benchmark 17-Mile grade is 2.2%. The Eckhart Branch experience provided the B&O with a not-to-exceed grade number. Luckily, the tonnage freight went downhill.

The company that delivered coal from the Maryland Mining Company to the B&O was Atkinson & Hazlehurst. They supplied the coal to the B&O for testing in locomotives in 1838. At that time, locomotives burned wood, or Pennsylvania hard coal. Coal went by wagon on the National Road. The Allegany coal trade kicked off in December 1843 when coal went from Eckhart to Cumberland by wagon on the National Road. There, it was loaded on the B&O to travel to dam #6 on the C&O canal west of Hancock, where it was transferred to canal boats for the trip to Georgetown. There, it was loaded on sailing ships for the journey to New York.

Maryland Mining was to use "immense locomotives of 25 tons weight, of B&O design and constructed by James Murray." Today's locomotives weigh in excess of 200 tons. Mr. Murray was Superintendent of Motive Power for the B&O, and approved of Ross Winans' *Camel* engine design, which were then being tested by the B&O for heavy haul in the mountains.

The Maryland Mining Company was also authorized to operate a bank in Cumberland, the Mineral Bank. The Company and its railroad were purchased by the Cumberland Coal & Iron Company in 1852.

The B&O Railroad provided early motive power and rolling stock to the Allegany County coal shortlines. The B&O supplied at least eight Camel engines to the Maryland Mining Company, as evidenced by Winans' notes, now at the Maryland Historical Society in Baltimore. These locomotives included B&O engines 161, 162, and 163, among others. In addition, Winans, among other builders, sold engines, tenders, and coal cars directly to the various mining companies. Passenger service was provided on the Eckhart Branch sometime before 1853, and the C&P continued to use a gravity passenger car on that line. The car was released in Eckhart, and allowed to roll downhill to Cumberland, under the control of a brakeman. The passenger car was later hauled back up the mountain at the end of a string of empty coal hoppers. Servicing, watering, and coaling facilities for the locomotives were located in Eckhart.

At the opening ceremony of the rail line on Wednesday, May 13, 1846, a special train took the board of directors and guests from Cumberland to Eckhart, and returned. About two weeks later, an accident occurred on the line near the junction with the Mount Savage Rail Road, at the west end of the Narrows. A dozen passengers were injured when the brakes burned out on the train, and it overturned due to excessive speed. It was noted in a contemporary newspaper account that these were the same brakes commonly used on the Baltimore & Ohio line, but they were not adequate for the steeper grades of the Eckhart Branch. Flooding in July of 1846 also caused extensive damage to the line's lower end, where Braddock Run meets Will's Creek.

The Potomac Wharf Branch was built by the Maryland Mining Company between 1846 and 1850, as an extension to the Eckhart Branch Railroad. The Potomac Wharf Branch crossed Will's Creek at Cumberland on a bridge (no longer present) just east of the current Route 40 road bridge. Some of the rail may still be seen. The area near the creek end of present-day Will's Creek Avenue is known as City Junction, and had a water tank and a tower. The Potomac Wharf Branch was crossed by the George's Creek & Cumberland Line. Rail was removed from the section west of the Valley Street crossing in Cumberland as late as 1990. In 1994, rail was removed from this area to maintain the Western Maryland Scenic Railroad to Frostburg.

A classic wreck scene photo, circa 1860, shows the bridge collapsed into Will's Creek, with engine *C. E. Detmold* dangling into the creek. The locomotive was named for the operator of the Lonaconing Iron Furnace. Extensive flooding of Will's Creek had caused the bridge pillars to give way. This image indicates the branch and the facility were in use at least to this date. The original Potomac Wharf Branch bridge was a 203-foot deck plate girder structure, with two support pillars in the creek. Built in 1849, and rebuilt after the Detmold accident, it survived until the flood of 1936.

Metropolitan Southern Railroad

This Baltimore & Ohio Railroad entity was established to build a spur line from Silver Spring, MD, to the Potomac and the Canal. In 1892, the company laid two miles of track, and built a large wooden trestle over Rock Creek, which was the largest such trestle on the system at the time. The line was used to supply coal to the electric power generating station of the Rock Creek Railway, a trolley line.

Metropolitan Western Railroad

This was a Baltimore & Ohio Railroad entity chartered in Virginia to build a rail line from the Potomac River to Quantico, using a new Potomac River Bridge north of Chain Bridge. Financing did not materialize, and the charter expired on March 1, 1899.

Mount Savage Rail Road

The Maryland & New York Iron & Coal Co. was charted in 1838, and built several blast furnaces at Mount Savage, Allegany County, Maryland. These furnaces were modeled on the George's Creek Coal & Iron Company's Lonaconing Furnace. Besides the blast furnaces, facilities were built to work the cast iron, most notably a rolling mill, where the first American made iron rail was manufactured in 1844. Five hundred tons of rail were produced for the company's railroad, which followed the path of Jennings Run to Will's Creek, and then through the Narrows to the B&O

railhead at Cumberland. Bridges were originally built of timber, but were later replaced by iron due to excessive maintenance costs.

In February 1844, records indicate that the B&O railroad supplied engines and cars to the Maryland & New York Iron & Coal Company. The 10 mile long Mount Savage Rail Road was completed to Cumberland in 1845, the same year Florida was admitted as a state. The bridge over Will's Creek west of the Narrows towards Mount Savage was dated 1842. The B&O did not send their best equipment into mine service. Engines of the "second class" were used. This classification was based on weight and performance, not quality. To put it in better terms,

"It must be remarked that the duty of the 2nd class engines appears so much less than that of the other classes, not from inferior efficiency, but from circumstances which have given the two engines of this class less to do than they could have accomplished. This is particularly to be said of the engine of this class which has done the work of the Mount Savage Road; this engine being, in fact, one of the best in the service." (21st. Annual Report of the B&O, Oct. 1847, p. 43)

In a single week in December 1852, the Mount Savage Rail Road moved more than 5,500 tons of coal to Cumberland. All of this had been dug by hand. The first commercial contract by the B&O Railroad to move coal was signed in February 1844 with the Mount Savage company. The B&O hauled coal to Baltimore, and some to Dam 6 near Hancock for reloading on canal boats. The railroad was a customer of the product for its own locomotives as well.

April 1, 1845, marked the date of an historic agreement between the B&O Railroad and the Maryland & New York Iron & Coal Co. It stipulated a charge of 1 1/3 cents per ton-mile to transport coal from Cumberland to Baltimore, provided the company shipped at least 175 tons/day for at least 300 days of the year. Coal was still viewed as a speculative commodity by the B&O. Wood or charcoal were the fuel of choice for industry, and for home heating. Connection was made with the B&O in Cumberland in 1846. Also in that year, the B&O contracted for 15 miles of Mount Savage rail, nearly 675 tons of the 51 pounds per yard product then produced. The rail was used to upgrade the line between Harper's Ferry and Baltimore. Before this purchase, the B&O was relying totally on imported British rail.

Getting coal from the mines into canal boats, and, earlier, into the Potomac River boats before the canal was completed to Cumberland, was a challenge. Along the North bank of Will's Creek, coal from the mines would be dumped on the ground, awaiting the spring floods that would allow boat traffic on the Potomac. At the right time, flat-bottom boats would be loaded by hand shoveling for the trip downstream. There was a wharf that the boats came alongside of, for loading. Coal made the trip from the mines originally via wagon on the National Road. This valuable piece of land along Will's Creek was owned by the Cumberland Basin Company, chartered in the State of Maryland in 1849. In 1850, the coal wharf on Will's Creek was the only such facility in Cumberland, and served exclusively by the Mount Savage Railroad.

Initially, canal boats could enter the Potomac River through the guard lock, and proceed upriver for some distance. The dam in the Potomac below the guard locks ensured that the Potomac was deeper at its junction with Will's Creek than it is today. The guard locks and the dam were removed as part of the Will's Creek flood control project, built by the Army Corps of Engineers for Cumberland in the 1950s.

B&O's Cumberland viaduct was built as a brick arch structure during the period 1849 to 1851. The Wharf Branch line and the B&O main passed through the "Deep Cut." The cut (passage) provides the "West End" of the B&O with access to the Potomac River Valley, towards Keyser, and Grafton. The viaduct passes over city streets, Will's Creek, and the WM tracks (ex-GC&C,

now used by the Western Maryland Scenic Railroad). The viaduct was originally single-tracked, but later a second track was added to it and the "Deep Cut" to the south. The southern end of the cut is wide enough for triple track, and the bridges are designed for three tracks. The C&P line merged into the B&O westernmost tracks, then crossed over to the easternmost track to access the boat loading wharf in the Potomac.

In 1845, the railroad inaugurated passenger service from Mount Savage, with connections to the B&O in Cumberland. Three trains per day were provided and operated by the B&O. At that time, the trip from Baltimore took 8 1/2 hours. William Cullen Bryant wrote of his trip in the Saturday Evening Post, providing a fascinating glimpse into the rigors of the early travels. He writes, "At Cumberland, you leave the B&O railroad, and enter a single passenger car at the end of a long row of empty coal wagons, which are slowly dragged up a rocky pass beside a shallow stream into the coal regions of the Alleghenies."

Norfolk Southern

Modern day class 1 railroad Norfolk Southern is a conglomerate of the previous Norfolk & Western, and the Southern Railroad Systems. It joined with arch-rival CSX Corporation to acquire the assets of Conrail in 1998, consolidating railroading east of the Mississippi into just two parts. Norfolk Southern Corporation was formed June 1, 1982, and operates over 21,000 miles of track in some 22 states in the eastern part of the U.S. Its primary cargo includes coal and intermodal traffic. It was formed from a series of predecessor railroads dated back to the some of the earliest railroads in the United States. It absorbed many of the smaller lines that crossed and interacted with the canal. Norfolk southern predominately operates south of the Potomac (and the canal), but does interchange with CSX at spots in Maryland.

Norfolk & Western RR

The Norfolk and Western Railway was the product of more than 200 railroad mergers spanning a century and a half. Beginning in 1838 with a 10-mile line from Petersburg to City Point, Va., NW grew to a system serving 14 states and a province of Canada on more than 7,000 miles of track.

In 1881, the AM&O (Atlantic, Mississippi, and Ohio Railroad of Virginia) was purchased by E.W. Clark and Co., a private banking firm in Philadelphia, and was renamed Norfolk and Western Railroad. Frederick J. Kimball, a partner in the Clark firm, headed the new line and consolidated it with the Shenandoah Valley Railroad. He and his board of directors selected a small Virginia village called Big Lick, later renamed Roanoke, as the junction for the Shenandoah and Norfolk & Western lines. The rail line crossed the C&O canal and the Potomac at Shepardstown, near lock 38 and the river lock. Served by the N&W, the Port of Norfolk was and is in direct competition with the Port of Baltimore for the coal export trade.

Pennsylvania Railroad in Maryland

The entity Pennsylvania Railroad in Maryland was incorporated on January 12, 1876, and the first rail was laid in May of 1879. When completed in June of 1879, the line stretched some 6 1/2 miles from Cumberland along Will's Creek to Ellerslie, MD, at the Pennsylvania state line. Here it connected with the Bedford & Bridgeport Railroad, controlled by the Pennsylvania Railroad. The first train ran on December 2, 1879. Eventually the railroad provided through passenger connections to Washington and New York. Coal traveled from the George's Creek region to Philadelphia, and the Pennsylvania Railroad's export dock at South Amboy, NJ. The Pennsy wanted to build their own line to Cumberland to bypass the C&P, since that company was controlled by Consolidation Coal, then owned by the B&O.

At the West end of Cumberland, an area known as City Junction, the line had to cross the C&P's Potomac Wharf Branch, which was there first. The Pennsylvania Railroad in Maryland line had been built from the Pennsylvania state line to the west side of the Narrows. It was their intent to continue down the north side of the Narrows, along with the C&P and B&O mains, to Cumberland. The C&P persuaded the other road to bridge Will's Creek, and continue down the south side of the Narrows, then cross the C&P's Potomac Wharf Branch at City Junction. When the line was built to City Junction, the C&P changed its mind. It kept an engine parked at the intended crossing point, blocking construction. When the engine was a bit late getting back into position one day, the Pennsylvania trackmen forced the crossing. The C&P trackmen tore it out. Tempers flared. The C&P raised its trackbed, making crossing impossible. The final issues were decided in court, in favor of allowing the crossing.

The Pennsylvania Railroad in Maryland was intended to provide competition to the B&O. The City of Cumberland held a $65,000 second mortgage, with no interest, for 30 years. This line was consolidated with the George's Creek & Cumberland in 1888. That line was absorbed into the Western Maryland Railway. Pennsylvania Railroad trains provided service to and from Cumberland until 1934. The Penn Central abandoned the line to Bedford in 1972, and the Western Maryland *State Line Branch* was removed in 1982.

Piedmont & Cumberland Railroad

The Piedmont & Cumberland Railroad (sometimes, Piedmont & Cumberland) was an operating entity of Senator Davis' West Virginia Central & Pittsburg Railway, formed in Maryland in 1886. It was absorbed into his West Virginia Central & Pittsburg Railway later. Senator Davis had been a B&O Station Agent at Piedmont, and the bounty of timber and coal resources of West Virginia did not escape his view.

His house in Piedmont still stands. The West Virginia Central and Pittsburg (WVC&P) began as a narrow gauge line in 1880, but changed its name and gauge in 1881.

The P&C was authorized to build a line in Maryland opposite the WVaC&P above Piedmont, to Cumberland. The rail line was opened to Cumberland from the south in 1887. This had the major advantage (for Davis' West Virginia coal) of approaching the canal wharfs from the south without requiring a crossing of the Baltimore & Ohio tracks, for which the B&O extracted a fee. The WVaC&P and all its subsidiaries were later sold to the Gould-controlled Fuller syndicate in 1902. They were merged into the Western Maryland Railway in 1905.

One of Senator Davis' business partners was his cousin, Senator Arthur Gorman of Maryland. If you look at the towns along the line of the WVaC&P south through West Virginia, you see the names of Davis' business partners - Thomas, Davis, Kerens, Gorman (and Gormania). Gorman influenced canal affairs for many years as a major political boss in Maryland Politics. He also became President of the Canal. Around 1889, when the canal was in deep financial difficulty, Davis and Associates were maneuvering to buy the right of way to allow access from the West Virginia coal fields by rail to the eastern ports.

Pittsburgh and Connellsville Railroad

The Pittsburgh and Connellsville Railroad was incorporated in 1837, and connected Connellsville and West Newton, PA, by 1855. The final link between Port Perry and Pittsburgh was not made until 1861. To facilitate construction and ensure its place on the line, Connellsville Borough voted $100,000 in aid to the railroad. Connellsville was a major supplier of coke to the steel mills at Pittsburgh. Although coke could be transported by barge on the river, it was faster to transport it by rail. The steel mills had a voracious appetite for coke.

This line was bought by the Baltimore & Ohio Railroad in 1871. This solved their problem of the interference by Pennsylvania Railroad, the B&O's rival, trying to keep it out of Pennsylvania. The line went from the canal basin at Cumberland following Will's Creek through the Narrows, and to the Pennsylvania State line, then west to Pittsburgh. You can ride this line today west on Amtrak's Capitol Limited out of Cumberland. It essentially follows the path of the proposed northern extension of the C&O Canal to Pittsburgh.

Potomac and Piedmont Coal and Railroad Company

The Potomac and Piedmont Coal and Railroad Company was founded by Henry Gassaway Davis in 1866. His business partners were T. B. Davis, W. B. Davis, W. J. Armstrong, J. Phillip Roman (attorney in Cumberland), R. G. Rieman and James Boyce of Baltimore. In 1880, construction began on a rail line from the interchange with the Baltimore & Ohio Railroad at Bloomington south along the Potomac. In 1881, the line reached Elk Garden, WV. In 1881, Davis reorganized the company in the States of Maryland and West Virginia as the West Virginia Central and Pittsburg. This would be the beginning of a line to feed West Virginia coal to the canal and Pennsylvania Railroad at Cumberland.

Salisbury and Baltimore Railroad and Coal Company

This line resulted from a name change in 1871 of the earlier Elk Lick Coal, Lumber, & Iron Company, chartered in Pennsylvania in 1868. John Anspach, of Philadelphia, was the key player. The line was sold at foreclosure in 1875, and then reorganized as the Salisbury Railroad Company. It became part of a consolidation in 1912 to form the Baltimore & Ohio Railroad in Pennsylvania. Their line went from a station, one mile west of Myers' Mills (Myersville, PA) on the Pittsburgh and Connellsville Railroad, extending to Salisbury (PA). No rail had been laid by December of 1873. This line followed the proposed right-of-way of the Will's

Creek (Northern) extension of the canal to Pittsburgh.

Shenandoah Valley Railroad

See section on the Cumberland Valley Railroad. This line became part of the Norfolk and Western.

South Branch Valley RR

The shops and operating headquarters of the South Branch Valley Railroad (SBVR) are in Moorefield, WV. The line, ex-B&O, runs south from the Potomac and the Canal at Greenspring, WV to Petersburg, WV, 52 miles, along the South Branch of the Potomac River. The line is operated by the West Virginia State Railway Maintenance Authority. It provides freight service to a series of industries along its length, including a charcoal plant and a chicken hatchery.

The Potomac Eagle is an excursion passenger service over SBVR tracks. It runs on weekends, and features a magnificent ride through the canyon of the South Branch heading towards the junction with the North Branch of the Potomac. A particularly scenic area is called the Trough, where nesting eagles can be seen.

Twin Mountain & Potomac

(TM&P – "two mules and a pony") was a short-lived West Virginia shortline. It was a 36" gauge line, stretching from an interchange east of the New Creek bridge at Keyser, WV south some 26 miles to Twin Mountain near Burlington. It was built to service the apple orchards in the area of Patterson Creek Mountain. These orchards were operated by Will & Allen Russel, and E.A. Letherman. Although the primary cargo was freight, the TM&P also handled passenger traffic, particularly church meeting specials. A circus special train was the source of a ballad about the railroad. Mixed freight was often run, with a TM&P box car and a passenger car behind one of the engines. The railroad operated from 1912-1919, and was scrapped by the B&O. Had the canal extended past Cumberland as far as Keyser, the TM&P could have provided cargo's of apples to the boats, and certainly to the hard-working mules. True to its name, the TM&P was built with mule power.

Washington & Cumberland Railroad

This Maryland company was incorporated by the State legislature in 1890. It was specifically formed to acquire the right-of-way of the C&O Canal as a railroad route from Cumberland to Georgetown. Although a separate corporation, the Washington & Cumberland (W&C) was seen as a proxy for Davis' West Virginia Central & Pittsburgh (WVaC&P) Railway. The Company's charter was quickly amended by the state legislature to "...have full power and authority to construct, complete, maintain, equip, and operate by steam or any other power, a railroad with one or more tracks upon and along the tow-path or bed of the Chesapeake and Ohio Canal..." It was also authorized "...to construct, equip, maintain, and operate branch or lateral railroads through the counties of Montgomery, Howard, Anne Arundel, and Baltimore to any point within the City of Baltimore." The amendment also stated the Corporation "shall at no time be owned or controlled by any railroad company owning or operating a competing parallel or nearly parallel road." i.e., the Baltimore & Ohio. At the same time, the Maryland Legislature passed an Act authorizing the C&O Canal to lease the canal to the W&C, and to waive its liens on the State in favor of the new company. The deal was all tied up in a bright ribbon. The WVaC&P would get the canal right of way, including the Paw Paw tunnel, and get to operate a competing railroad, and the B&O could not do anything about it. The State of Maryland would be gratefully out of the canal business. A message from Maryland Governor Jackson to the Maryland Legislature urged acceptance of the proposal. A bill was passed in the District of Columbia allowing the new railroad access to the City.

This was not the first attempt by the State of Maryland to dispose of its holdings in the canal to a railroad. In 1895, it was reported that the State of Maryland had put out for bid its interest in the Canal. Three bids were received. One was from the B&O Railroad (then in receivership); one from Davis' WVaC&P Railway, and the third was from Richard Kerens of St. Louis, a business partner of Davis. All three bids were rejected. In 1888, a bill was presented to the Maryland Legislature to allow the Western Maryland Railway to lease the canal for the purposes of constructing a rail line. This scheme was also connected to Davis' WVaC&P Railway, and supported by Hood, the President of the Western Maryland Railway, and by the ex-President of the Canal (who just happened to be Arthur Gorman, a business partner of Davis). The bill did not pass. Davis had Charles H. Latrobe do a survey for the proposed line. Latrobe was the son of Benjamin Latrobe, Jr., the Chief Engineer of the Baltimore & Ohio Railroad, and architect of many of its impressive civil engineering works. Surveying and preliminary grading was done in anticipation of the project.

The W&C was formed by Enoch Pratt of Baltimore, David L. Barrett, John A. Hambleton of a major Baltimore banking house, Asa Willison, a City Councilor of Cumberland, Battersby W. Talbott of Montgomery County MD, and E. K. Johnson of Washington. Hambleton was a business partner of Henry Gassaway Davis, of the WVaC&P Railway. The Washington & Cumberland was compelled by its charter to "...run at least two passenger and two freight trains daily in both directions upon said railroad for the whole distance thereof, unless prevented by floods..." History along the canal showed the disclaimer to be necessary.

As reported in the New York Times, February 6, 1890, the W&C Company bid on the canal, proposing a lease of 99 years, with a plan to use the canal as a right-of-way for a railroad to Georgetown. Also, the rail line would branch off at Williamsport to Baltimore, using Western Maryland Railway track. It was also authorized to interchange with the Frederick & Pennsylvania Railroad, and the planned Baltimore, and Drum Point Railroad, which reached the Chesapeake Bay near Lusby, MD. This would open up the West Virginia coal fields directly to the coastal trade, in a path not involving the Baltimore & Ohio Railroad. It was never completed due to financing problems, and the expansion of the Port of Baltimore.

The Baltimore & Ohio Railroad, not surprisingly, opposed the plan. The other Canal bondholders (other than the State of Maryland) opposed the deal as well. The lease deal, carefully orchestrated by the Democratic State Machine, fell through. The Canal Trustees had to post a bond to put the canal in working order, and operate it profitably. They had until 1891 to make this happen. Having exited from receivership, the Baltimore & Ohio Railroad paid the bond. The B&O did not want the canal, or to operate the canal, but it also did not want the canal to fall into the hands of a rival. The State of Maryland and the Maryland Board of Public Works was not happy about the situation. By 1891, the B&O was firmly in control of the canal. Davis shut down the Washington & Cumberland project in exchange for $190,000 from the B&O, and a "favorable agreement on rates" from the B&O and the Pennsylvania Railroads.

Washington & Old Dominion Railroad

The W&OD began in 1847 as the Alexandria, Loudoun & Hampshire Railroad in hopes of competing with the Chesapeake & Ohio Canal. It was in operation by the Civil War. Plans to cross the Alleghenies never materialized and rails only reached Bluemont, VA, a resort town on the eastern slope of the Blue Ridge Mountains. The line became the Washington & Ohio Railroad in 1868 and added "Western" to its name in 1883. Until 1912, it ran trains into D.C. Union Station. By 1912, it became the Washington & Old Dominion, was electrified, and moved its passenger terminal to Georgetown. Trains crossed the Potomac and the Canal on a former aqueduct bridge. A branch to Great Falls, VA, was operated until abandoned when the amusement park there closed. At the end, passenger equipment was usually a diesel-powered combine car which took 2 hours and 45 minutes to go from

Rosslyn to Purcellville. In 1956, The Chesapeake & Ohio Railroad bought the W&OD to serve a proposed power plant, which was never built. The railroad did well hauling material during the construction of Dulles Airport, but was shut down later, and the right-of-way abandoned.

Washington & Western Maryland Railroad

This was a Baltimore & Ohio Railroad entity created to build a spur line from Potomac Palisades to Georgetown in 1906-07. It would have crossed the Canal, but was never built.

Western Maryland Railroad

It has been said that the reason for the existence of the Western Maryland Railroad was to export all of the state of West Virginia, particularly the mineral wealth and timber. The Western Maryland Railway started out in the 1850s as an agricultural railroad, operating in central Maryland. It serviced the areas that the B&O didn't. Chartered by the Maryland Legislature in May of 1852 as the Baltimore, Carroll & Frederick Railroad, the name was changed in 1853. The company incorporated the Baltimore & Susquehanna Branch from Relay House to Owings Mills in 1857. New construction headed towards Westminster. As they did for the Baltimore & Ohio Railroad, the City of Baltimore provided a bond guarantee for the construction. The charter was to construct a railroad from Baltimore to Westminster, and then towards Hagerstown. This would open up Carroll County and the Cumberland Valley region in Washington County to the markets of Baltimore. Construction was interrupted by and began again after the Civil War. In 1873, the line had reached Williamsport, west of Hagerstown, and the key connection with the Chesapeake & Ohio (C&O) Canal. Coal was shipped on the canal to Mr. Victor Cushwa at Williamsport, for the use of the Western Maryland Railroad. Coal also went by wagon north to Hagerstown and Pennsylvania. The Western Maryland provided a link between the canal at Big Pool and Baltimore. A unique lift bridge at MP 99.5

allowed the WM access across the canal to the power plant, for coal shipments. It was built in the winter of 1923-24, and is still in place.

In 1902, the City of Baltimore divested itself of its WM bonds to the Fuller Syndicate, controlled by Gould and Rockefeller. The Gould master plan involved a transcontinental railroad, built as a consortium using existing connector roads and shortlines where possible. It was a battle of the robber-barons, with the Pennsylvania Railroad, the B&O, and the New York lines fighting for the freight traffic of the nation. The line from Big Pool was immediately extended westward to Cumberland to connect with another piece of the Gould puzzle, the West Virginia Central & Pittsburg. (The name of the Town later had the h added to the end). The Western Maryland Railroad (Railway, after Dec. 1, 1909) reached Cumberland, MD, from the east in 1906. West of Big Pool, it required 23 bridges and five tunnels (Indigo, Stickpile, Kessler, Welton, and Knobley), following the water level route of the Potomac River. The line was closed in 1975., and the right-of-way is being used as a rail-trail.

Overextended financially, and outmaneuvered by its rivals, the Gould-headed syndicate, and the Western Maryland Railroad, went into receivership in 1908. The railroad emerged in 1910, reorganized as the Western Maryland Railway. The Connellsville extension was completed in 1912. The current station in Cumberland was built in 1913. During World War I, the WM was directed to oversee operations of the Cumberland & Pennsylvania Railroad. The government oversight of the railroads lasted from 1917 to 1920. The WM was doing well, after surviving almost being dragged down with the Gould Empire.

In 1927, the Baltimore & Ohio purchased John D. Rockefeller's Western Maryland stock. This gave it a 43% interest in its rival. The Interstate Commerce Commission (I.C.C.) told the B&O to divest the stock, but, as a compromise, the stock was put in trust with Chase Bank in New York. This was the beginning of the end for WM independence. The WM completed numerous purchases and consolidations in the ensuing years, including the Cumberland & Pennsylvania in 1944. Passenger service on the WM continued south out of Cumberland to Elkins, WV, until the 1950's.

In the area of the Paw Paw bends, the canal snakes along the Potomac and through its tunnel. The Western Maryland Railway and the Baltimore & Ohio Railroad straighten out their paths with a series of trestles and tunnels. At South Cumberland, the Western Maryland crossed the B&O line, and then the canal and the Potomac to use Welton tunnel through Knobley Mountain in West Virginia to reach Ridgeley and the yards and shops at Maryland Junction. The Western Maryland Elkins subdivision was the old West Virginia Central and Pittsburgh line south to West Virginia. The bridge leading to Welton tunnel can be seen crossing the Potomac south of CanalPlace.

The Western Maryland Railway wanted to buy enough of the canal right-of-way from Williamsport to Cumberland to allow it to extend its way west to the coal fields. That never happened, so it built a roughly parallel line to accomplish the same thing. After consolidation with the B&O, the WM line was abandoned as redundant to the B&O line south of the Potomac. The WM line is now a rail-trail.

In 1904, the State of Maryland was disposing of its holdings in the canal for whatever price it could get. The Western Maryland Railway bid $155,000 for $5 million in stock, and additional loan rights, and won the bid. With the Gould Syndicate failure in 1907 (Gould owned the WM at the time), the stock ended up with the B&O (which was controlled at the time by the Pennsy). The railroad business was cut-throat.

Western Maryland Scenic Railroad

The Western Maryland Station in Cumberland is located adjacent to the old guard lock and inlet of the C&O Canal. The Station houses the National Park Service Canal office, a transportation museum, and the ticket office for the Western Maryland Scenic Railroad, which provides a steam powered ride along WM and C&P right of way to Frostburg.

The WMSR tourist excursion operates daily steam or diesel-powered trips to Frostburg, Maryland. The line includes portions of the Western Maryland's Connellsville extension, originally built as part of the George's Creek & Cumberland Railroad in 1913, and of the Cumberland & Pennsylvania. Both lines were absorbed into the WM.

Operations began on the WMSR in 1989, with steam motive power provided by Pacific type (2-6-2) steam engines of the Allegany Central, and a city-owned GP9 diesel. The City of Cumberland diesel was painted black, and lettered in near-WM speed lettering style with a road number 40. It was a GP9 high-nose unit, ex-C&O 5940. It went back to Virginia with the Allegany Central. Later, diesel units were leased from Sheridan Rail Operations, including an Alco RS3 and an RS-D5 unit. Three VIA rail FA2 units were obtained, and two were painted in WM and one in B&O colors. One of the units in WM colors (No. 306) was later sold to the South Branch Valley Railroad in West Virginia, and has since been repainted. Unit 305, the other unit painted in WM colors, was also sold. FA unit 800, in B&O colors, went to the Liberty Limited Dinner Train operation out of New Freedom, PA along with the Western Maryland lettered RS-D5.

The Scenic Railroad is operated with City of Cumberland, Frostburg, Allegany County, and State of Maryland funding. The Allegany County tax on motel rooms goes to support this tourist magnet. The money has been well-spent, as more and more visitors come to ride the train, then stay to shop, dine, and visit other attractions in the area.

The line extends some 15.6 miles from Cumberland to Frostburg, with a maximum grade of 2.8 percent. There are four bridges and one tunnel on the line. The first bridge, just to the west of the Cumberland Station, takes the line across Will's Creek. The next bridge, at the west end of the Narrows, carries the line over Route 40, Braddock's Run, and the old Eckhart Branch Rail Road right of way. The next two bridges are on ex-C&P trackage. They once carried the C&P line over the Western Maryland tracks towards Connellsville. Brush Tunnel, the only one on the line, was built by the Western Maryland, and is 834 feet long.

The pride of the Western Maryland Scenic motive power fleet is a 2-8-0 steam engine, painted WM Fireball, and numbered 734. It is visually similar to a WM class H8 or H9 Consolidation model from Baldwin. It may sometimes be seen operating in conjunction with one of the diesels, in a scene reminiscent of the early 1950s. The current WMSR diesels, ex-Conrail GP30's, were painted in sports themes at various times. One was decorated for the Washington Redskins football team, later changed to the Baltimore Ravens, and the other was decorated in Baltimore Orioles baseball colors. Both now wear Western Maryland Railroad colors. As of Fall, 2024, the WMSR has a new steam locomotive in the shops, C&O 1309. This is from the B&O Museum in Baltimore, and is being put back into running order.

The Cumberland Canal Basin, now CanalPlace, is behind Cumberland's Western Maryland Station, used by the Western Maryland Scenic Railroad. When the re-watering project of the canal is completed, it will be possible to transfer between rail and canal boat at Cumberland with a short walk.

West Virginia Central & Pittsburgh Railroad

The West Virginia Central & Pittsburg Railway was chartered in Maryland and West Virginia by Henry Gassaway Davis. He was a B&O Freight agent at Piedmont, and saw the value of transportation systems to the vast mineral wealth and timber of what was then western Virginia. He started the Potomac and Piedmont Coal and Railroad Company in 1866. He then built a rail line from a junction on the Baltimore & Ohio Railroad near Bloomington, heading south to the coal fields of Elk Garden, along the north branch of the Potomac. The company name was then changed to WVaC&P. The rail line was extended to Fairfax, WV in 1884, the site of the future town of Davis. This was a logging, sawmill and tannery center, and home to the Davis Coal & Coke Company. In 1886, the rail line built north from the B&O Junction towards Westernport with a goal of Cumberland. This was done under the Piedmont and Cumberland Railroad Company name, later the "WVaC&P in Maryland." Cumberland was reached in 1887, and connections were made with the canal and the Pennsylvania Railroad in Maryland. The yards in Cumberland were located west of the Baltimore Street Bridge across Will's Creek. Locomotives and rolling stock were purchased from the Cumberland & Pennsylvania. The Cumberland Central Station, along Baltimore Street at Will's Creek, was the terminus of this line. The WVaC&P now needed a connection with the canal.

Construction continued in West Virginia with the rail line going south to Parsons (named for a business partner of Elkins) in 1888, and Elkins (named for his son-in-law) in 1889. Elkins was a major railroad hub. A line ran north and west along the Tygart River to Belington. Another line ran along the river south to Beverly and Huttonsville.

As the financial condition of the canal worsened, the West Virginia Central had its eye on purchasing the canal, to use it as a ready-made right-of-way for a rail line. Arthur Gorman was President of the Canal Company and a Director of the WVaC&P when the railroad began buying up newly-issued 1878 canal repair bonds. The value of the canal lay in the 184 mile right-of-way, just ideal for railroad tracks. The Baltimore & Ohio Railroad did not want any further competition, and helped to block this scheme. The Maryland Board of Public Works wanted to recover the State's investment in the canal, many millions of dollars. The Board had made many attempts to sell its canal stake, all unsuccessful. In 1885, it advertised for bids, and received three. One was from the

Washington & Cumberland Railroad (a Davis entity), one from the WVaC&P, and one for a small amount of cash. All bids were rejected. In 1889, the Board tried again, but only received a small cash offer from the B&O. On another try, the WVaC&P bid was rejected again. In 1905, a bid was finally accepted from the B&O, for less than half of the amount of smaller bids having been rejected before. The B&O preserved the canal from those who wanted to transform it into a rail line, and the State was off the hook, although it did not come close to recovering its investment.

The Fuller Syndicate bought the WVaC&P and associated railroads in 1902, and the system was merged into the Western Maryland Railway. Parts of the WVaC&P line are still in operation. Some of the line is used by CSX for coal traffic, and the West Virginia State Rail Authority has a tourist rail operation in Elkins, Belington, and Durbin, WV. In the same area is the Cass Scenic Railroad, a former logging line. But, the canal remained more-or-less full of water, not paved over and not tracked.

Williamsport, Nessle, and Martinsburg

This line crossed the Potomac and the Canal on a bridge about a mile east of Dam 5. The bridge served until the flood of 1936, when it was swept away. The bridge piers are still visible in the river, and one sits next to the towpath. This line ran from Charlton, MD, to Marlowe, WV near the West Virginia line, and became the Charlton Branch of the West Subdivision of the Western Maryland Railway. From the state line to the end of the rails at Synder, it was known as the Williamsport, Nessle, and Martinsburg Railroad. This was, however, a wholly owned subsidiary of the Western Maryland, and operated with WM equipment. Its purpose was to move fluxing limestone from Virginia to the Carnegie Steel Plant in Pittsburg, which it did from 1915 to 1931.

Winchester and Potomac Railroad

The Winchester and Potomac Railroad (W&P) ran 28 miles from

Winchester, VA, to the B&O junction at Harper' Ferry. It opened for service in 1836, and was absorbed by the Baltimore & Ohio in 1848. It was extended 20 miles to Strasburg, Virginia to connect with the Strasburg and Harrisonburg Branch of the Virginia Midland Railroad. The rich agricultural output of the Shenandoah Valley was tapped. The W&P Railroad was eventually absorbed into the CSX system. It played a key role in early Confederate raids on the B&O during the beginning months of the Civil War. The line was used by Stonewall Jackson to transfer the captured machinery of the Federal Arsenal at Harper's Ferry south to Winchester.

When the newly formed B&O planned to cut across the northern end of the lower Shenandoah Valley, the W&P Railroad was chartered by the Virginia Assembly in 1831. Routes were surveyed by the U.S. Army Topographical Engineers from 1831 to 1832. The B&O was completed to Harpers Ferry in 1834, and the W&PRR was completed by 1836. The railroad began operations on March 14. A Bollman truss bridge connected the north end of the Winchester and Potomac Railroad to the B & O towards the end of the Civil War

The railroad terminated at the corner of Water and Market Streets in Winchester. The Winchester depot immediately became a key economic hub serving merchant traders in Winchester for commodities such as wheat, hide, fur, tobacco and hemp. The north end of the rail line also served the thriving industrial town of Virginius Island, which sat astride the Shenandoah Canal on the south side of Harpers Ferry.

Winchester & Western Railroad

The Winchester & Western Railroad crosses the canal and the Potomac River east of Williamsport, and proceeds north to Hagerstown. It is still operating in daily service, moving sand from a quarry in Gore, VA. This is a ways from its original (1916) purpose in being built, which was to tap the Virginia hardwood forests for lumber for railroad ties. It shared a station with the Baltimore & Ohio in Winchester.

East to west; Tidewater to the mountains

This section takes a look at key canal and railroad intersections, industries, and infrastructure along the canal route, heading west from Georgetown. Distances along the canal are measured from Georgetown as MP - mile post, following Hahn's Canal Guide. For more in-depth information, see Hahn's excellent towpath guide. Unlike Hahn, I am not going to discuss every location along the line. For more information on the named railroads, see the previous chapter. After we discuss the situation at Cumberland, we'll talk about some of the compare-and-contrast issues with the various modes of transportation, the canal, the railroads, and the National Road. Right now we have to head up stream, locking through, and drawn by a hard-working team of sturdy mules. Don't be in a hurry.

Georgetown (MP 0.0)

Georgetown was incorporated as a Maryland Town in 1789. It was a major seaport at the time, but later the Potomac silted up, and ocean going traffic stopped at the Port of Alexandria, VA. This also thwarted the British Navy in the War of 1812, who wanted to knock out the Foxall Foundry in Georgetown, the primary source of American cannon. Georgetown was the site of many grain and cotton mills, and these were operated by water power. Georgetown was about as far as you could go upriver on the Potomac, with the Little Falls and the Great Falls providing a major obstacle.

By the 1820s, the waterfront silted up and was navigable only by smaller coastal vessels. . Construction of the C&O Canal began in July 1828, to link Georgetown to Harper's Ferry, Virginia (now West Virginia). The canal was completed to Cumberland on October 10, 1850. The canal turned out not to be profitable, never living up to expectations. The Canal did provide an economic boost for Georgetown. In the 1820s and 1830s, Georgetown was an important shipping and industrial center. Tobacco and other goods were transferred between the canal and shipping on the Potomac

River. Salt was imported from Europe and sugar and molasses were imported from the islands of the West Indies. These products were later superseded by coal and flour, which flourished with the C & O Canal serving not only as a transportation avenue, but also providing cheap power for mills and other industries.

Georgetown's transportation importance was defined by its location just below the fall line of the Potomac River The Aqueduct Bridge (and later, the Francis Scott Key Bridge) connected Georgetown with Northern Virginia. Previously, a ferry service owned by John Mason connected Georgetown to Virginia. In 1788, a bridge was constructed over Rock Creek to connect Bridge Street (now known as M Street) with the Federal City. Alexandra, Virginia, was the deep water port for distribution.

The Alexandria Canal was built to connect the C&O Canal at Georgetown with the Port city on the Potomac. In Georgetown, Wisconsin Avenue is built on the alignment of the old tobacco rolling road from rural Maryland and the Federal Customs House (now the post office) was located on 31st Street The city's oldest bridge, the sandstone bridge which carries Wisconsin Avenue over the C&O Canal dates to 1831. It is the only remaining bridge of five constructed in Georgetown by the Chesapeake & Ohio Canal Company.

Georgetown was a busy place at the lower end of the canal, and delays in unloading caused traffic backups. Finally in 1877, an inclined plane was built upstream to allow boats not unloading at Georgetown to bypass the four lower locks and directly enter the Potomac. It was independently funded, and paid for by tolls. It had three sets of railroad tracks. In the center, a large caisson held a canal boat in water, and counter weights were on the outer two tracks. It was water-turbine powered to provide the additional energy required to lift the caisson.

Several streetcar line and interurban railways interchanged passengers in Georgetown. The trolley station was located in front of the stone wall on Canal Road (currently occupied by a gas station) adjacent to the steps made famous by the movie "The Exorcist," and the former DC Transit trolley car barn at the end of the Key Bridge. Four suburban Virginia lines, connecting through Rosslyn, VA, provided links from the D.C. streetcar network to destinations such as Mount Vernon, Falls Church, Great Falls, Fairfax, Vienna, Leesburg, and Purcellville, VA. Streetcar operations in Washington, D.C. ended January 28, 1962, superseded by buses. The B&O Railroad built a branch line from Silver Spring in Maryland to Water Street in Georgetown in an attempt to construct a southern connection to Alexandria. It served as an industrial line, shipping coal to a power plant on K Street until 1985. The abandoned right-of-way has since been converted into the Capital Crescent Trail, a rails-to-trails route. The power plant is now a condo.

Commerce and industry developed along the waterfront, where wharves and flour mills were constructed. During the Revolution, Georgetown served as a great depot for the collection and shipment of military supplies. When the town was finally incorporated in 1789, a textile mill, paper factory, and more flour mills were already established there.

Point of Rocks (MP 48)

Site of a picturesque E. F. Baldwin-designed B&O station, circa 1875, Point of Rocks is where the B&O's Old Main Line (from Baltimore) and Metropolitan Branch (from DC) join and head westward. It is located at milepost 42.8 on the B&O Metropolitan Sub, where the Catoctin Mountains reach the Potomac. The area saw much action during the Civil War. The site was originally called Washington Junction. Here, KG tower, located south of the Metropolitan Branch, controlled rail traffic from 1905-1959. At Point of Rocks, the B&O Railroad first ran into contention with the C&O Canal, then being built west from Washington. The situation did not improve for the railroad until they bored the nearby Catoctin Tunnel in 1868. Alternately, stone for the canal locks was brought in by rail.

Brunswick (MP 53-54)

Brunswick was the site of major B&O freight yards and a roundhouse. The area is now used for MARC engine refueling and cleaning. The B&O had major classification yards and servicing facilities here. In the 1890's, they could handle 4,000 cars. In 1906-1907, the west yards were added, with an additional capacity of 4,250 cars. The Brunswick turntable was long enough to handle B&O's heaviest locomotives. The 1910/1917 19-stall B&O roundhouse was torn down in 1995. Brunswick was a both a railroad and a canal town. The town was founded in 1780 as Berlin, called Barry Post Office, and then Brunswick. It was occupied by Confederate forces for a time during the 1862 Antietam Campaign in the Civil War, and again in 1863, disrupting rail traffic on the B&O. Brunswick got a passenger station in 1879. The name was changed to Brunswick by the B&O Railroad, not by the Post Office.

The current rail station is a 1891 E. F. Baldwin design, and has been moved from a previous site in town. The memorials to the crew and passengers of the MARC passenger train wreck of 1995 are located adjacent to the station. The Brunswick Railroad Museum is located at 40 W. Potomac St.

Harpers Ferry (MP 60.7-61)

Harper's Ferry has a long history of important events, including John Brown's raid on the Federal Arsenal, and Stonewall Jackson's capture of numerous B&O engines and rolling stock during the Civil War. The Harper's Ferry rail station, located in the old town, is used for Amtrak service. At Harper's Ferry, the Shenandoah River flows into the Potomac. The C&O Canal also passes along the north bank of the Potomac, forcing the railroad into a narrow strip of land along the cliffs. The B&O line passes through an 1894 tunnel, and immediately over the Potomac River Bridge. Several bridges have crossed the Potomac over the years, becoming victims of Potomac floods and Confederate action. The B&O River crossing is on the National Register of Historic Places. Most of the Town is a National Historical Park.

Robert Harper, an Englishman, was granted a land patent of 125 acres in the area in 1750. He took over a ferry crossing across the Potomac in 1761. This opened up the Shenandoah Valley from the North. The town was established in 1763. Washington visited the area in 1785 to examine the area for potential bypass canals, like had been used at Great Falls. He proposed the site as appropriate for a National Armory and Arsenal. The area became a major industrial center before the Civil War, producing muskets, rifles, and pistols. John Brown's raid on the arsenal was perhaps the most famous occurrence in the town, which changed hands eight times during the Civil War. The Federal troops mostly destroyed the strategic arsenal and its machinery, and the Confederate forces salvaged and removed what was left. Due to its strategic position, the town was heavily garrisoned and heavily damaged during the war.

A wooden frame train station built in 1894 was moved in 1931. The station had an integral tower for control of the B&O's Shenandoah Branch, previously the Winchester & Potomac Railroad. The current rail alignment is roughly over the site of the old Federal Arsenal. MARC and Amtrak passenger trains serve the Town, and the old B&O line is now heavily used by CSX freight service.

There was once an Amusement park built by the Baltimore and Ohio Railroad on Byrnes Island. Such projects were used by the railroads to encourage tourist travel on the line

The manufacturing Island of Virginius was adjacent to Harper's Ferry, along the Shenandoah shore. It hosted a series of water-driven mills for manufacturing. In fact, from inlets on the Shenandoah side, a series of underground water distribution system were used to power turbines in the basements of the buildings. Some facilities used water wheels.

Antietam (mp 69)

Antietam was the site of the bloodiest battle of the Civil War, but was also a large industrial site at one time. The industrial village extended up Antietam Creek. There were three lime kilns, and an iron furnace. The site was active from the early 18th century and had deposits of iron ore, limestone, and a plentiful supply of hardwood for charcoal. In 1750, the area was known as Frederick Forge. The furnace began operations in 1765, and produced pig iron that was used to cast cannon and ball for the Revolutionary War. Craftsmen also fashioned muskets. Product was shipped south on the Potomac. Later, pig iron was supplied to the new Federal Arsenal at Harpers Ferry. The production was 40-60 tons per week. Coke replaced charcoal as the preferred fuel for the furnaces in the 1860's, and coal delivery from Western Maryland became cost-effective.

The industrial site hosted a large number of mills, driven by a raceway from Antietam Creek. A water-powered bellows for the iron furnace was also built, as well as a water-powered 21 ton forge hammer. The adjacent rolling mill was also operated by water power. The water-powered nail factory produced 500 kegs per week. In 1840, the iron works were said to employ 260 men.

Shepherdstown (MP 72.2)

This Potomac River and C&O Canal town in Virginia (later, West Virginia) was chartered in 1762. Water power is abundant in the Town. More than six natural springs feed Town Run before it enters the Potomac. The creek never floods, nor runs dry; and it meanders through backyards, under houses, across alleys and beneath streets. This setting was conducive to millers, tanners, potters, smiths and other artisans. In addition, the water power and transportation artery of the river and later the canal was nearby. It was in the river at Shepherdstown that Rumsey demonstrated to

Washington a steam-powered boat that could go upstream. Washington wrote, "...it might be to the greatest possible utility in inland navigation..." The C&O Canal was built across the river from the Town in 1830, and the Shepherdstown Lock is number 38. Most of the town is on the National Register of Historic Places.

Williamsport (MP 99)

This town could have become the Nation's Capitol. Congress had stipulated that the new Federal City had to be built somewhere between the Anacostia on the east, and Williamsport on the west. The citizens spiffed up the Town up for George Washington's inspection tour. The final site for the Federal City was chosen to be near the Anacostia River by George Washington. This was partially based on easier access for ocean-going traffic, as the Nation's Capitol was also envisioned as a major center of commerce.

Williamsport was reached by the canal in 1835. The Western Maryland Railroad reached the town from Baltimore in 1873. Local businessman Victor Cushwa and others were merchants and canal shippers, dealing in coal, plaster, cement, & fertilizers. They built warehouses and docks at the basin. Before the coming of the railroad, merchandise could head north to Hagerstown and Pennsylvania in freight wagons.
Williamsport was a key point for trade to the north and a major intermodal point on the canal and the railroad. Here, canal boats, two railroads, and numerous horse-drawn wagons converged.

Green Spring, MD (MP 110)

At Green Spring (MP 110) there was an iron furnace. The coming of the canal triggered the construction of a second furnace in 1848. Nine hundred tons of pig iron went to Georgetown in one year, probably to DuVall's and Foxall's foundries. Fort Frederick, at MP 112, had been built by Maryland in 1755-56 to serve as a fall-back point if Fort Cumberland was taken by the French in the mid-1700's.

Big Pool (MP 113)

At Big Pool there was a Western Maryland Railroad station and a canal basin. The Western Maryland had a bridge across the Potomac at Cherry Run to connect with the Baltimore & Ohio. Cherry Run also had a train station and railroad office building. Some canal families wintered over at Big Pool.

Hancock (MP 124)

Hancock at Mile 124 is a canal and railroad town. It now hosts a Canal Visitors Center. It was the home of Moffet Station, on the Western Maryland Railroad; the rail line reached town in 1905 from Baltimore. Unfortunately, the station burned in 1982. There was a B&O Station across the river on the West Virginia side. Today, the Western Maryland Rail-Trail, runs 10 miles from Hancock east to Fort Frederick, and extends west along the old WM right-of-way towards Cumberland.

The Round Top Cement Mill near MP 127 produced hydraulic cement from Round Top Hill. The facility began in 1837 as Shafer's Cement Mill. The Round Top Cement Company was formed in 1863. The product went east and west to be used in canal construction; the area also produced sand for glass and bark for tanners. In 1882, the company employed as many as 100 men, and had an associated cooperage shop to produce the necessary barrels. There were eight cement kilns, using coal from Cumberland. Water power ran the grinding mill. The mill capacity was 2,200 barrels per week. Shipments were by canal, and the B&O Railroad.

Benjamin Mitchell, boat builder, had his business in Hancock. At Dam 6, the B&O transferred some coal from rail to canal boat from 1842-1850 until the canal was completed to Cumberland.

Little Orleans (MP 131)

Little Orleans was on the main route from Fort Frederick to Fort Cumberland in colonial days. It became a major shipping point for lumber after 1850. Docks were built along the canal banks and a ford in the river allowed a connection with the B&O Railroad across the river at Orleans (VA). Products could be shipped east or west. In the early 1800s, the town of Orleans Crossroads VA grew very rapidly after the arrival of the B&O Railroad. On the Maryland side, the town was smaller-and used the name "Little" Orleans. The Western Maryland Railway, stretching along the Maryland side of the Potomac, reached the town in 1904. It built a station and water tank in the town. The Billmeyer Lumber operation had a siding at Little Orleans until 1927. In addition, a barrel making business had a small shop in Little Orleans, which shipped out their product via rail.

Paw Paw (MP 156)

Located in West Virginia, the small town of Paw Paw gave its name to the only tunnel on the C&O Canal. The Town is named for a local fruit. In the vicinity, there are similar railroad tunnels on both the Western Maryland and the Baltimore & Ohio railroad lines. Due to the meanders of the Potomac, sometimes the rail lines emerge from one tunnel, cross the river on a trestle, and immediately enter another tunnel. The topological challenges were the same for the canal and the railroads. The Green Ridge Railroad had a connection with C&O Canal at Darkey's Lock (lock 67) across from Paw Paw, WV, on the Maryland side of the Potomac.

The C&O tunnel is 3,118 feet long and a single boat wide, a choke point on the route. It was estimated to take 2 years to construct, but it required 14, with a 4 year hiatus.. It overran its construction estimate of $600,000 by a factor of twenty. The required number of tunnels in the area resulted in a labor shortage of skilled and unskilled labor. The original Baltimore & Ohio Railroad line is in active service by CSX Corporation and Amtrak. The old Western

Maryland Railway right-of-way is being converted to a hiker-biker trail.

The Town of Paw Paw does have another treasure besides the tunnel, the B&O Railroad Station, an E. F. Baldwin design.

Mertens Sawmill; Green Ridge Railroad (MP 160)

The Green Ridge Railroad belonged to the Mertens family of Cumberland, who owned extensive timber lands in Green Ridge. The rail line hauled timber to a sawmill at Oldtown, for use by the Merten's boatyards in Cumberland to construct the canal boats. The railroad operated from 1889 to 1897. It also interchanged with the B&O Railroad across the Potomac at Okonoko, WV, after crossing the river and canal on a trestle.

Town Creek Aqueduct (MP 162)

Adjacent to the Canal's Town Creek Aqueduct there is a Western Maryland Railway trestle bridge. The access road to the aqueduct area runs on the old WM right of way. From the aqueduct, looking across the Potomac, one can see the current CSX tracks, ex-B&O. The trestle is in bad shape, and heavily overgrown.

The South Branch of the Potomac flows north into the North Branch Junction near MP 165. The South Branch Valley Railroad was built by the B&O to tap the agricultural products of the South Branch Valley. It still operates under the West Virginia Railway Authority.

Oldtown, MD (MP 166.7-167)

This was the area where early pioneer Michael Cresap established a fortified house, and engaged in the Indian trade. It was a stopping point for Washington, and for the Braddock Expedition against the French at fort Duquesne. This area saw exciting activity during the Civil War, with a Union armored train defending against

Confederate forces, but being disabled by cannon fire. This site has two canal locks fairly close together. The Merten's sawmill was once located here.

Patterson Creek (MP 173.5)

An auxiliary steam pump at Patterson Creek supplied some 100-120,000 gallons of water per minute from the Potomac River. Low water levels in the River, particularly in the summer months, made the steam pumping operation a necessity for the canal. The pump probably used coal delivered by canal boat.

Western Terminus – Cumberland, MD, Milepost 184.5

Cumberland was predestined by geography to become a transportation nexus, where various rail lines, the Chesapeake & Ohio (C&O) canal, and the National Road would come together. The Baltimore & Ohio (B&O) Railroad's goal was to push westward quickly to tap the lucrative grain trade of the Ohio Valley. Cumberland was just an intermediate goal. The coal fields of the nearby George's Creek Region proved a huge traffic source that continues to today. Local industries thrived, serving not only the transportation needs, but providing export goods to world markets. These industries shaped and defined the region's railroads. Numerous significant railroad related activities happened in the Cumberland area in the 19th Century. The first iron rail manufactured in the United States was rolled at Mt. Savage, a few miles from Cumberland. Also at Mt. Savage, the shops of the Cumberland & Pennsylvania Railroad turned out their unique steam engines for their own use, and for sales to other lines. At Frostburg, another manufacturer, T. H. Paul & Sons turned out narrow gauge mining and logging railroad equipment.

In the early 1700's, if you traveled up the Potomac River from Tidewater, you would reach the confluence of Will's Creek in a few days ride. To the West, Will's Creek emerges from an impressive canyon, now called the Narrows. There was a small

Native American settlement, and the Ohio Company built a small wooden blockhouse and trading center on the south side of the Potomac. . The lands belonged to Lord Fairfax of Virginia, or to Cecil Calvert of Maryland; no one has yet done the survey. More importantly, the area is on the westernmost border of the British Empire in North America with New France. The American part of the European 7-years war would be fought here. The Native people fought on both sides of the conflict, and, in the end, were the losers.

The City of Cumberland was founded in 1787, but before then, Fort Cumberland was the westernmost outpost of the British Empire in North America. West of Fort Cumberland, towards the Ohio River, was New France. Fort Cumberland had been built by Virginia and North Carolina colonial militia. From here, the Braddock expedition against the French at Fort Duquesne was launched, and ended in disaster near the present day location of Pittsburgh. The frontier could not be held, and was pulled back to the vicinity of Frederick, Maryland, where a substantial stone fort had been built.

After the French and Indian War, and the American Revolution, President Washington returned to the Fort to review the troops in the Whiskey Rebellion incident. By this time, a small settlement has built up along the river and the creek.

Cumberland is a transportation nexus, defined by geography and geology. The National Road heads west from there. The Baltimore & Ohio railroad and the C&O Canal raced to Cumberland on their way west to the Ohio River. The second airport in the nation was built there, half-way between the Wright Brothers factory and the first airport at College Park, MD.

Based on coal-driven technology, the 19th century economy boomed. No major actions took place during the Civil War, sparing the area most of the ravages of war. A bored Union General, in encampment, found time to work on his novel, "Ben Hur."

Political sympathies were mixed. The Town was mostly occupied by Union Forces, except for a brief interval when the Mayor was obliged to surrender the Town to Confederate forces.

After the war, the industrial revolution really took over, as major and minor industries sprang up. These included glass factories, railroad shops, canal boatyards, iron foundries, a cotton mill, and their supporting industries. The homes of the rich and famous lined fashionable Washington Street.

Cumberland, MD, was never intended as the western terminus of the canal, it was just the point where the canal company ran out of money. The canal had been foreseen to transport vast quantities of grain from the vast western lands beyond the Ohio River to the eastern seaboard. This will be explored in more detail in a later section. Even before the canal or the railroad got to Cumberland, coal was transported in flat bottom boats to Georgetown in the spring when the river was high. It was stockpiled along Will's Creek in anticipation of the high water. Cheap boats were built in Cumberland for the transport; they didn't make the return journey, but were sold at the end of the line. The intrepid boatmen walked back. Some polled lighter boats upriver, taking close to a month for the return journey. Empty boats could be brought back upstream, but loaded boats were next to impossible. Not much cargo was worth the effort.

If we trace what actually came upstream and up-canal, we see a shift from Cumberland as a frontier outpost, to a community with a need for manufactured goods, to a manufacturing center. Technology and manufacturing went west, enabled by the transportation,

Cumberland, as the western terminus of the canal, developed an infrastructure to support National Road, railroad, and canal operations. In anticipation of the arrival of the canal, various businesses were being set up to exploit the new opportunities. The Cumberland and Georgetown Transportation Company was chartered in the State of Maryland in 1853 by William A. Bradley, Charles M. Thruston, and Richard S. Cox. Bradley was mayor of Washington D. C. from 1834-1836, and was the Cashier of the Bank of Washington. He was also the Postmaster. He owned the island now known as Roosevelt Island in the Potomac. General Charles M. Thruston was a graduate of West Point, who moved to Cumberland in 1837. His son was a representative to the Maryland General Assembly, and States Attorney for Allegany County. Richard S. Cox was a Treasury official, living in Georgetown. He had served as paymaster for the Confederate Army, and his home was confiscated by the government. It is now the Duke Ellington School for the Arts. He and his family moved back to Georgetown after the war.

The National Road Era

For a period of 20 to 30 years, the National Road became the preferred avenue for moving people and goods. This required the construction and repair of wagons and stages and tack. The system required methods of horse maintenance, veterinarians, farriers, and feeding stations. A system of change-out stations was required every few miles. This grew into a series of taverns (for the stagecoach trade) and the rougher roadhouses (for the teamsters). Many teamsters and horses were needed. Barrels were required to ship products such as fish, cement, and whiskey. The National Road construction was financed by the Federal Government by sale of public lands in Ohio. It was built 64 feet wide, and cost $13,000 per mile. The construction cost of the National Road, from Cumberland to Wheeling on the Ohio River, was $1.7 million. Another million was used to complete the toll road to Cumberland from Baltimore. Soon, mail coaches were taking advantage of the new pathway. On the National Road, not all cargo traveled in wagons, some of it was self-mobile. Vast herds of cattle, pigs, and sheep were driven along the road, snarling traffic for the faster and more organized coaches and wagons.

Even after the railroad reached Cumberland, the stagecoaches and freight wagons completed the journey to Wheeling. There was active cooperation between the railroad and the various stage and freight lines to provide through service. Before the turnpike was completed, moving a ton of freight from Baltimore to Wheeling cost $120. That same ton of freight, shipped from Liverpool to Baltimore cost $12 to deliver. When the National Road was opened, the price dropped to $35/ton. Later, the railroad brought freight from Baltimore to Cumberland in 30 hours for $6/ton. It then went by wagon an additional 6 days to reach Wheeling at an additional $10/ton. In 1835, the all-wagon trip would have taken 2 weeks. The all-rail trip was feasible by 1853. This changed the economics of shipping. Things that had been too expensive to ship were becoming available.

Boatyards

Cumberland was where almost all of the C&O Canal boats were made. In the 1880's there were at least seven firms involved in the construction and maintenance of canal boats. The basin area first contained William Ward's boatyard until 1853. It then became Weld's. In 1884, it became Weld & Sherridan. Other firms included Doener and Bender, William Young, the Consolidation Coal Company, R. and M. Coulehan, Isaac Gruber, and W.A. Meredith & Company.

The Weld & Sheridan Boat Building and Repair Yard (1884-92) was located where the current Crescent Lawn at CanalPlace area is today. The lay of the land is very different than it was when the boat yard was active. The 1950's flood control project in Will's Creek by the Army Corps of Engineers made major changes to the landscape. A downstream dam in the Potomac River was removed in 1954 to lower the water levels, and Will's Creek was channelized. The canal basin had been unused for many years, before being filled in in the 1920's and 30's.

Henry Thomas Weld was an English immigrant. The Weld's Lulworth Estate is located in central south Dorset, England. Its most notable landscape feature includes a five mile stretch of coastline on the Jurassic coast. Part of the area is a special World Heritage Site. The estate is predominantly owned by the Weld family who have lived there for multiple generations. The Lulworth estate was once part of a grand estate under Thomas Howard, 3rd Viscount Howard of Bindon. The historic estate, of which the stately Lulworth Castle, which was the residence to the Weld family until 1929 when it was ravaged by fire. Henry was associated with Weld's Boatyard in Cumberland, later the Weld & Sheridan Boat Building & Repair Yard at the C&O Canal basin. Captain John Sheridan was mentioned in conjunction with the Union Army in 1864, as Assistant Quartermaster at Steubenville, Ohio.

A C&O Canal boat cost about $2000, a good boat-wright making $3.00 per day in 1880. C&O canal boats were constructed using oak for structural members, and pine for decking and planking. Square beams were joined with butt joints. Composite beams for the bow and stern allowed for the curves. The maximum size of the boat was set by the lock size, 14 feet wide and 90 feet long.

The facilities uncovered during an excavation of the area in 2004 included a saw pit and a marine railway. Multiple intact hulls and part of hulls of canal boats were found buried in the area, lying parallel to each other. At least seven boats were found and documented, before being reburied. Funds for recovery and restoration were not available. The boats and other artifacts await future generations with more money and better techniques.

An interesting story appeared in the Cumberland paper after the report was issued. An old fellow living on a farm east of Cumberland read the article in the paper about the excavation, and came into town to see. Turns out, he had lived on a canal boat as a young child. He watched the boats being sunk and the basin filled in. He knew all that, if someone had just asked him.

Mertens Boatyard

The Mertens boatyards in Cumberland were major construction and repair yards for the many canal boats. They required a constant source of wood, and this was provided by the Mertens sawmill in Oldtown. This was in turn fed by the Mertens forestry holding on Green Ridge, east of Cumberland. Logs came down to the Potomac River on the Green Ridge Railroad, a Mertens-owned narrow gauge shortline, using locomotives built by T.H. Paul & Sons in Frostburg, MD The lumber was hauled into Cumberland on the B&O Railroad, or on the canal boats heading back upstream. The boatyard site is now called CanalPlace, and is located behind the Western Maryland Railway Station. The Crescent Lawn Archaeological District is located within CanalPlace, and is Maryland's first Certified Heritage area. The Merten's boatyard operated until around 1905. In 1892, the Queen City Foundry was also located in the area.

The Western Maryland Railway Station was built near the location of the canal basin in 1906 and this also impacted the facilities and layout of the canal. Later construction of the Footer's Dye works also changed the land. The business was founded in the 1870's, and became the largest business of its kind in the country by the turn of the century. The dye works dumped their waste water into the canal basin. A single building of the Footer Complex survives at CanalPlace.

Railroad Connections

When the B&O arrived in Cumberland in 1842, there were several short-line railroads already in place or in the process of being built. These included the railroad of the Mount Savage Coal & Iron Company with its Mount Savage Iron Works.

The B&O Railroad provided motive power and rolling stock to the Allegany County coal shortlines. There was a discussion in the early days of railroading about who should own the freight cars, the railroad or the shipper.

The railroad facilities included the passenger depots and the servicing facilities. Freight warehouses for rail car to freight wagon transfer were mundane. Coal was loaded at the mine site, directly into the rail cars. Passenger stations became a major part of the community.

When the B&O had reached Cumberland, it basically obsoleted the stage coach lines that ran from Baltimore. However, stage-to-rail connections in Cumberland provided through passage to Wheeling before the railroad was extended that far. Stages north into Pennsylvania and south to Virginia were also available, feeder lines to the railroad.

Boat wharfs

There were a series of boat wharves at Cumberland for coal loading. The Mount Savage Rail Road built one in 1850 along Will's Creek. This was the property of the Canal Basin Company. The area had been used for years for coal delivery. The earliest activity involved coal arriving in horse-drawn wagons on the National Road, and dumped on the ground. There were a series of wharfs where flat-bottomed river boats could be tied up, and loaded by hand. These boats could reach Georgetown when the river level was right; especially, in the Spring. As the canal proceeded westward, these boats could enter the canal via river locks and inlets until the westernmost point was completed, and use that pathway to continue their trip to Georgetown.

The Potomac Wharf Branch of the Eckhart Rail Road was built by the Maryland Mining Company between 1846 and 1850, as an extension to their Eckhart Railroad. This allowed coal delivery by rail to the Potomac Wharf in the Potomac River, upstream of the entry point of Will's Creek. The wharf was built by John Galloway Lynn, a Cumberland businessman. His heirs sold this facility to the Maryland Mining Company.

The Lynn's were a prominent 18th century Maryland family. Lynn's father had moved to Cumberland, building a substantial brick house on the West Side known as Rose Hill. In 1849, Lynn and the Mount Savage Rail Road incorporated the Cumberland and Pittsburg (sic) Rail Road Company. Their eyes were on the grade over the Alleghenies, but nothing seems to have come of the venture. Lynn was also an incorporator of the Lulworth Iron Company in Maryland, active in Mount Savage.

According to noted rail historian and photographer William P. Price, the rail siding was one thousand feet long, extending from present day Kelly Blvd. around to the mouth of Will's Creek. The Potomac Wharf was listed on the C&P's Interstate Commerce Commission valuation sheets in 1918, although it is doubtful it was in use at that time. Coal loading was probably done by side-dumping coal from rail cars through chutes directly into the canal boats. Much easier than shoveling.

The best and final approach to coal loading was the canal wharfs, along the canal to the east of the Potomac and Will's Creek, the current location of CanalPlace. The construction of the flood control channel for Will's Creek has drastically altered the landscape in this area. The wooden Canal wharf allowed a railcar of coal to be dumped directly down into a canal boat. The facility was only strong enough for one carload of coal, and locomotives were not allowed on the structure. A horse pulled the car to the proper location. Why a horse and not a mule is not recorded. The loading process was in a first come, first served type of arrangement, depending on how well you knew the wharf master and coal brokers. Boats were also re-provisioned and necessary repairs and veterinary services were available for the boat's motive power at the Canal Basin. There were also a few saloons.

The 1923 Interstate Commerce Commission valuation docket for the C&P (in the National Archives) gives the construction date for a substantial concrete wharf as 1917. The B&O provided access for the C&P to reach the Canal Wharf, charging a tonnage tariff for this access.

Railroad Facilities

After the B&O arrived at Cumberland in 1842, rail support infrastructure was needed to support the operations. The B&O built a roundhouse downtown to service locomotives. A second roundhouse was built in 1896, a mile or so east of the original one. This facility was later moved to South Cumberland, with a new 32-stall roundhouse for the locomotives. It is still in use at this writing by CSX Corporation. Large classification yards for freight cars are also located here. The locomotives and freight cars were built in Baltimore. Flour moved in barrels carried in boxcars. As coal began to become the dominant freight traffic, special coal cars were developed. Improvements in equipment such as car couplers and brakes lead to increased efficiency and reduction in accidents and injuries.

There were four daily trains between Baltimore and Cumberland initially, a passenger and a freight in each direction. The line was single-tracked, with a passing siding every 8 miles or so. The railroad was run by timetable, and, in most cases, successfully.

The B&O's Queen City Station and Hotel was built in 1871. It was destined to last for one hundred years. It combined 150 guest rooms, formal gardens, a ballroom, kitchens, laundries, facilities for crew, and a telegraph office. Across the tracks, the B&O had an express facility and freight offices.

The Bolt & Forge facility was built by the B&O in 1870's with a grant of land by the City of Cumberland. It was intended to recycle scrap iron for additional railroad use, and to forge custom parts. It produced a large quantity of steel rail, and grew to be a large facility. The site is now a shopping center.

The B&O YMCA, a home-away from home for many of the crew,

was built in south Cumberland, across Virginia Avenue from the B&O Roundhouse. Later, the B&O engineer's school was located near the south Cumberland facilities.

WVaC&P

The West Virginia Central and Pittsburg entered Cumberland from across the Potomac at Ridgeley, WV, currently the maintenance facility of the Western Maryland Scenic Railroad. It connected with the Pennsylvania Railroad in Maryland to provide service to Harrisburg and Philadelphia. It later interchanged with the B&O and the canal, bringing coal from West Virginia. Yard facilities were located on the east bank of Will's Creek, upstream of the Baltimore Street Bridge. Later, extensive shop facilities and rail yards would be built in Ridgeley West Virginia.

GC&C

The George's Creek and Cumberland Railroad had its facilities at the east throat of the Narrows in Cumberland, where Mechanic and Centre Streets come together. There was a ten stall roundhouse and extensive shops. No part of this facility has survived.

Consolidation Coal Company

The Consolidation Coal Company was chartered in 1860. It was allowed to "transport persons and property on the railroad at the same rates of toll as the B&O (ref. Charters, Acts of Legislation, and By-laws relating to the Consolidation Coal Company and the Cumberland and Pennsylvania Railroad Company of Maryland, 1872, New York, William R. Vidal, Stationer.)." The president and directors had the power to "purchase and place on the railroad all machines, wagons, vehicles, or carriages of any description whatever..." In 1864, it purchased the stock of the Cumberland & Pennsylvania from the Mt. Savage Iron Company. Consolidation Coal's ownership of the C&P continued until the Western

Maryland Railway takeover. Large amounts of Consol stock were acquired in 1875 by the B&O. Robert Garrett was named to the Consol Board in 1876. Garrett was the legendary president of the Baltimore & Ohio Railroad; the westernmost county in Maryland is named after him. Consolidation Coal didn't just own mines and railroads. The coal company had their own fleet of 30 canal boats, and a series of coastal schooners. The company owned coal wharves at Cumberland, Georgetown, Alexandria, and Locust Point, Baltimore. They were getting into the total vertically integrated intermodal package, coal from source to customer.

Cumberland Coal & Iron Company

The Cumberland Coal & Iron (CC&I) Company, chartered in 1850, purchased the Maryland Mining Company's mines and railroad property, including the village of Eckhart, in April 1852. The rail line was extended to the nearby Hoffman mines in 1859. Cumberland Coal & Iron was in turn acquired by the Consolidation Coal Company in 1870._This company built the connection between the canal and the railroad for coal loading at Cumberland, in 1856.

Service on the Eckhart Railroad was hard, as evidenced by a series of correspondences with the Winans works in Baltimore in 1856, preserved at the Maryland Historical Society. On June 16, 1856, Cumberland Coal & Iron ordered a replacement right-hand cross-head for the engine *Braddock*. The *Braddock* had gone into service on July 1, 1854. On September 24, the same part was needed for the engine *Eckhart*. The *Eckhart* had been placed into service on August 1, 1849. A frantic telegram on December 9, 1856, emphasizes the need for urgency for shipment of the replacement left-hand cross-head for the *Eckhart*. The engines *Black Monster* and *Cumberland* were at work at that time. The parts were delivered to the B&O Railroad at Cumberland. It is not known whether the repair work was done at Cumberland, or at Eckhart. The engine *Eckhart* was later rebuilt at the C&P shops in Mount Savage in 1868.

Cumberland & Pennsylvania Railroad

The C&P acquired the facilities of the earlier Mount Savage Rail Road in Mount Savage, MD. This town is 12 miles or so west and north of Cumberland. The C&P built extensive shops facilities, and built their own locomotives after getting experience in repairing and enhancing models from other manufacturers. The shops facilities were established by James Millholland. Besides building locomotives for their own use, they went on to build for other railroads across the United States. They also built and repaired rail freight cars, engine tenders, and cabooses. A former master mechanic of the C&P, T. H. Paul, spun off a narrow gauge locomotive business, and located in Frostburg, MD, adjacent to the C&P passenger station. The C&P also maintained maintenance facilities in Eckhart, the shops of the Eckhart Rail Road that they had acquired, and at Franklin, near Westernport.

The original locomotive shop was constructed of stone and was 90 feet x 250 feet in size with a roof 33 feet high. An adjoining car shop, built at about the same time, was also of stone and was later extended with a wooden structure. These buildings still stand in Mount Savage as does the C&P enameled brick headquarters building. The Mount Savage Works sent two engines to the National Exposition of Railroad Appliances in Chicago in May of 1883. A locomotive catalog, issued by the Mount Savage Works, shows 5 basic types of locomotives for narrow gauge and main line use.

The C&P tended to make most of their own parts, only buying specialty items such as steam gauges and water injectors. At Paul's shop in Frostburg, many wooden patterns were kept for casting locomotive parts, but were unfortunately discarded. One of Paul's locomotives went on to work a sugar plantation in Cuba.

The C&P Locomotive Works experience shows the development of technology and the westward movement during the Industrial

Revolution in America. In 1850, the locomotives on the B&O were built in Baltimore, probably using Western Maryland coal. By 1856, major repairs were being made in Cumberland, using assemblies from Baltimore. By 1868, the Shops at Mount Savage were turning out complete locomotives.

Passenger traffic from Cumberland

The railroad was a transportation infrastructure; it enabled people in the outlying communities to go to market, and to attend school in the cities. Passenger through-service was provided by connection with the B&O at Cumberland and Piedmont. Combination tickets were popular. These provided round trip transportation and a ticket to a popular weekend picnic destination, and later for Chautauqua meetings.

The canal was never a big people-mover. The journey took too long, but was cheap. Packet Boats moved people and mail, but were limited in speed to avoid damage to the canal. You could get passage on a coal boat for cheap, probably for nothing if you agreed to help with the work.

Maryland Mining Company Rail Road

Passenger service was provided on the Maryland Mining Company Railroad to Eckhart sometime before 1853, and the C&P continued to use a gravity passenger car on that line. The car was released in Eckhart, and allowed to roll downhill to Cumberland, under the control of a brakeman. The passenger car was later hauled back up the mountain at the end of a string of empty coal hoppers. The schedule was adjusted to meet the Baltimore trains of the B&O at Cumberland.

At the opening ceremony of the rail line on Wednesday, May 13, 1846, a special train took the board of directors and guests from Cumberland to Eckhart, and returned. About two weeks later, an accident occurred on the line near the junction with the Mount Savage Rail Road, at the west end of the Narrows. A dozen passengers were injured when the brakes burned out on the train, and it overturned due to excessive speed. It was noted in a contemporary newspaper account that these were the same brakes commonly used on the Baltimore & Ohio line, but they were not adequate for the extreme grades of the Eckhart Branch.

In 1872, according to the schedules published in the Frostburg Mining Journals, there were two round trips a day from Cumberland to Eckhart. The ticket price was 37 ½ cents.

George's Creek and Cumberland Railroad

Passenger service was provided from Lonaconing to Cumberland on the GC&C. That line has a dedicated passenger locomotive. In 1887, there were two trains per day. If you took the 10:45 AM from Lonaconing, you could lunch in Cumberland before catching the 1:15 PM to New York over the Pennsylvania Railroad. This express service would arrive in New York at 7:10 AM the next day. There was checked baggage service on this line. The GC&C used the Hay Street Station in Cumberland, located along the B&O line near the viaduct. The B&O charged too much to other roads to use their Queen City Station.

Cumberland & Pennsylvania Railroad

The C&P used the Hay Street Station from 1878. It provided service to the George's Creek Valley. In 1872, according to the schedules published in the Frostburg Mining Journals, there were two round trips a day from Cumberland to Eckhart, and two from Cumberland to Piedmont. A Brill Gas-Electric car replaced steam-power for passenger service in April 1929. No diesel equipment was ever used by the C&P.

The C&P provided the mail and the railway express service to Frostburg and the mining communities of the George's Creek. The advantage for the railroad of carrying the mail was two-fold: it provided regular revenue that could subsidize passenger service, and it provided a level of protection from interference. With mail on board, it was known that interference with a train was 'a federal offense'. The railway express/mail car was usually placed first in the consist, behind the engine tender. The railroad charged per bag of mail, not per weight, so the bags were usually stuffed full. Bags of empty mail sacks, called "bums", rode for free. Will Lowdermilk, Postmaster of Cumberland, and author of the "History of Cumberland, Maryland" was responsible for giving the U.S. Mail contract to the C&P in 1872. Mail by train was faster than the alternative, which was the stagecoach. With multiple trains per day, good service was provided to the various mining communities of Eckhart, along the Jennings Run, and the George's Creek areas.

Baltimore & Ohio Railroad

The B&O featured passenger service between Baltimore and Cumberland from the time when the rails were first completed. The line was then single-tracked. The trip from Baltimore took 8 1/2 –10 hours at that time, which was a vast improvement over the stagecoach. By 1886, there were 8 passenger trains each way per day at Cumberland. Later, the B&O line would host up 20 passenger trains per day through Cumberland. The premier trains were numbers 1 & 2, the National Limited, Baltimore to St. Louis, and Trains 5 & 6, the Capital Limited, Baltimore to Chicago.

The first B&O depot was built along the tracks at Baltimore Street, adjacent to the Revere House Hotel. This was a transfer point to and from the various stage lines going further west. Locomotives were able to re-coal directly from the hopper cars of the Mount Savage Rail Road.

West Virginia Central and Pittsburg Railroad

The line between Cumberland and Elkins, WV, had passenger and combine cars as early as 1888. These were ordered from the major passenger car manufacturers of the time, such as Jackson & Sharp, and paid for with coal. They also bought from the C&P Shops in Mount Savage. The rail line eventually rostered a respectable 36 pieces of passenger equipment, including a business car. Emphasis on the line remained the coal shipments. In the 1920's, it was possible to book an Elkins, WV to New York trip, going via

Baltimore.

Western Maryland Railway

Western Maryland passenger service featured their K2-class Pacific locomotives, and green & gold passenger cars. Later, Alco RS-3 diesels replaced the steam locomotives. The WM Connellsville extension, built in 1911, called the "New Line," allowed for Baltimore to Chicago passenger service via Cumberland. Nicknamed the *Fast Freight Line,* passenger service was never a priority. Western Maryland's last passenger run east from Cumberland to Hagerstown was on May 30, 1953. The last passenger service out of the WM Cumberland Station was on January 4, 1958, to Elkins, WV. The Cumberland WM station now hosts the Western Maryland Scenic Railway tourist line, with steam excursion service to Frostburg, and the C&O Canal Museum.

Mount Savage Rail Road

By 1845, passenger service was provided to Mount Savage, using B&O equipment over the Mount Savage Rail Road. Three trains per day were operated, with one coordinating with the arrival of the Baltimore train.. There is a contemporary account of the train ride to Mount Savage from the Saturday Evening Post of 1860.

Pennsylvania Railroad in Maryland

Pennsylvania Railroad passenger service from Cumberland to Bedford began December 15, 1879, and was offered from the West Virginia Central and Pittsburgh Station at Baltimore Street and Will's Creek. It offered connections via Harrisburg to New York, DC, and Philadelphia. The mail was carried on this line. Service ended in 1936.

The trolley

In the 1890's, interest began to form in a trolley system to service the George's Creek area. In 1893, the Lonaconing and Cumberland Electric Railway was incorporated. This was followed by the incorporation of the Frostburg, Eckhart, and Cumberland, the Lonaconing, Midland, and Frostburg, and the Westernport & Lonaconing lines by 1901.

In 1901, construction started from Frostburg towards Cumberland. By 1902, the light rail line stretched from Frostburg down the George's Creek to Lonaconing. The first passenger run was made on April 24, 1902. At Cumberland, an interchange was made with the Cumberland Electric Railway, a local city service. There was a ticket office and terminus at Baltimore and Centre Streets. Hourly service was provided. The Cumberland and Westernport Electric Railway was formed by merger in 1906. Extension of the system to Salisbury, Pennsylvania, and to Keyser, West Virginia, were considered, but never built. Miner's specials ran down the George's Creek, to provide transportation for the different shifts.

The trolleys also carried the mail and parcels over their 27 miles of standard gauge track. The growth of freight and express service lead to the use of a freight-only trolley, making two trips per day.

Mostly Brill equipment was used, with some Southern cars being acquired later in the operation. There was a coal-burning 500-kilowatt power station and a car barn at Clarysville, serviced by the C&P. There was an auxiliary 400-kilowatt power station at Reynolds (near Westernport) for the Georges Creek line.

By 1924, the private automobile was making inroads on ridership of the traction lines. The trolley operation was sold to Cities Service, which replaced the trolleys with buses and freight trucks by 1925. Rail service was discontinued between Frostburg and Westernport on July 22, 1925 and between Frostburg and

Cumberland on August 4, 1926. The right-of-way and equipment was transferred to the Cumberland & Westernport Transit Company, which held the right-of-way until 1943. The Company was finally dissolved in 1955. County-provided bus service replaced the trolleys.

The C&WE was not to be outdone in computational capability. The *Cumberland Times* mentions, on March 6, 1916, "An ingenious machine – The Cumberland and Westernport Electric Railroad had just installed in their office a very ingenious machine that adds, subtracts, multiplies, divides, works proportion, cube root, and shows the position of the decimal."

Amtrak

The current waiting shed for Amtrak service, and a postal facility sit on the original site of the B&O's magnificent Queen City Station. This was a 150 room hotel with ornate gardens and fountains. The station was demolished in 1972, an act which spurred on conservation efforts for architecturally and historically significant structures, and for the preservation of railroad history.

A lot of interesting passenger equipment passes through Cumberland, including numerous examples of privately owned railcars (known as *Private Varnish*), Amtrak's experimental Turbo Train of the 1970's, and the luxury American-European Express. The Ringling Brothers, Barnum & Bailey Circus Train still passes through town on its way between major shows.

Daily passenger service through Cumberland is provided in both directions between Washington, DC and Pittsburgh, PA.

Relative Economics of Transportation Options

The C&O Canal began in 1828 and reached Cumberland in 1850. It was 184.5 miles long. It cost more than $11 million to construct according to Unrau, and was very costly to repair. It had to shut down in the winter due to freezing, and was subject to frequent damage from flooding. The Canal was financed by stock, and later, bonds, mostly purchased by the States, the City of Baltimore, and on the London Market. The cargo carrying capacity of the Canal was estimated to be 300,000 tons per year.

The B&O Railroad cost over $7.6 million to reach Cumberland (B&O Railroad, Annual Report, 1845). The path length was 179 miles. It started in 1828 and reached the Town in 1842. The B&O was a private company, and funds for construction were provided by bonds bought by the businessmen and City of Baltimore, and the States of Maryland and Virginia.

In 1828, the Railroad had capital stock to the extent of $4 million. Three million of this was from private individuals and corporations, with the rest from, equally, the State of Maryland and the City of Baltimore. The canal, on the other hand, had $3.6 million in capital, $3 million from public sources, and $600,000 from individuals. Of the public sources, $1 million each came from the Federal Government and the City of Washington, $500,000 from the State of Maryland, and $250,000 each from the Cities of Alexandria and Georgetown. This money in the form of bonds needed to be paid back, and was subject to interest. Debt servicing became an issue.

Later, in 1836, the State of Maryland chipped in another $3 million each to the canal and the railroad. The City of Baltimore provided an additional $3 million to the railroad. The railroad lead to Baltimore, and the canal did not. The methods of internal improvement in Maryland (the Canal, the Railroad, and the roads) lead to the development of a new creative financing method - preferred stock. In 1836, new stock issues were given preference over older investments for payment. Without this mechanism, new stock issues would have been worthless. With this mechanism, older stock holdings became devalued. The Maryland Legislature came up with the idea of preference stock. In addition, the new stock carried full voting rights. So Maryland put money into the projects, was first in line for repayment, and had effective control over the canal. Of course, the earlier investors had been passed over, but once the Ohio River was reached, there would be plenty of revenue for everyone. In reality, the Canal was hard pressed to meet operating expenses and repairs, let alone pay off its construction debts.

The cargo carrying capacity of the railroad was estimated to be over 500,000 tons per year. Cycle time for the round trip from the mines to Baltimore and back was estimated at 3 days. True to its name, the railroad did go from Baltimore, eventually to the Ohio River and beyond.

An issue encountered early on was the ownership of the freight cars. The B&O owned the locomotives, but were reluctant to put up the funds for the rolling stock. The customers were supposed to do this. Various mine owners bought cars for use in hauling their own coal. The problem was getting a strong, durable car with a low tare (empty) weight. You don't get paid to haul the empty. Ross Winans' 8-wheeled cars could carry three times their empty weight in coal. In this case, the canal boat had the advantage, because the weight was offset by the buoyancy. Canal boats, full or empty, required much less power to move. Freight cars were in short supply. The iron pot hoppers were replaced by wooden, and later, steel cars. Examples of these can be seen at the B&O Museum in Baltimore. The mining companies owned fleets of hopper cars, canal boats, and ocean-going sailing ships and steamers for the coastal and export trade.

All of the coal feeder shortlines in Western Maryland were built to the same gauge as the B & O, 4 feet, 8 ½ inches between the rails. This facilitated the interchange of rolling stock and motive power.

The railroad and the canal were in competition for the Shenandoah flour trade at Harper's Ferry in the early 1840's. From Harper's Ferry to Baltimore, the railroad charged 34 cents per barrel plus 3 more cents in fees. This was quickly raised to 50 cents. The railroad got the canal to raise its rate, then reduced its own. The flour on the canal went to the mills in Georgetown, not to Baltimore. Later, the battle for the coal trade from Cumberland overshadowed that of any other cargo. Through 1845, the canal and the railroad cooperated in the coal trade. The B&O hauled coal to Williamsport for trans-shipment down the canal, before the canal was completed to Cumberland. After the canal was up and operating, the agreement was terminated. The State of Maryland had an interest in seeing more product go to Baltimore. The railroad now instigated a rate war with the canal over coal transportation. In the 1870's, the railroad was hauling more coal than the canal.

Horses versus steam

It was demonstrated early in the Industrial Revolution in England that reducing friction was the key to moving large loads long distances. The development of canals using animal power used this principle, where only inertia has to be overcome to get the load moving, and the drag at low speeds is minimal. Go ride the replica canal boat today and see how easily a loaded boat is moved by one person.

On the ground, it was shown that "one horse (on a rail system) could do the work of eight on a common road." These early rails were made of wood, with flanged wheels on the cars, and were in use in the 1600's. Later, iron strap rail was used to reduce friction and extend the wear-life of the track.

The legal limit for horse drawn wagons on the roads of England was set at 36 bushels of coal, or about 3,000 lbs. of cargo. More than that would damage the roads too much. Roads were rutted and muddy, and hard-surfacing was in the future. McAdam introduced his process in 1820.

On unimproved trails, pack horses would have been used. The load has to be balanced on the two sides of the horse, and it can't exceed 30% of the horse's body weight. Horses (or mules) must be trained to operate in pack. Once that is done, they can function in the most unimproved terrain, and don't require a separate food source, except for grazing.

With heavier rail made from iron, steam engines were demonstrated as early as 1804. Not just to replace water power to drive mills, but as mobile locomotives to replace horses. Did this make economic sense? Absolutely. By 1814, steam locomotives were pulling 30 tons at 4 miles per hour upgrade.

The steam locomotive showed its worth at the Rainhill Trails in England in 1829. George Stephenson won the competition to supply locomotives. Only one horse-powered vehicle was entered, and it had the horse running on a treadmill. It did not work well. The Baltimore & Ohio Railroad started with horse-drawn equipment, but rapidly switched to steam power as Yankee ingenuity improved the British designs for American use. Before long, locomotives of British design were being turned out by American shops.

The use of steam propulsion on the canal was limited by several factors. Neither paddle wheels at the sides or stern of the ship were feasible, due to the size limitations of the locks. Screw propellers were tried, but churned up the mud in the canal prism and caused erosion. Steam navigation was limited to 4 mph, the speed of mules. Thus, there was no advantage to steam power. On the Potomac, however, steam boats were useful in heading upstream against the current. Some steam tugs were used in the canal, towing a "train" of barges. However, at each of the locks, the train had to be disconnected and the boats locked through one at a time. This was a lengthy process. Steam tugs were used at Cumberland to haul canal boats to the loading wharf in the Potomac River, before the loading facility at the canal basin was built.

A Captain J. L. Cathcart regularly plied a steam boat between Cumberland and Georgetown around 1858. Part of the advantage of steam navigation, he asserts, is that "...steam will cause an improvement in morals, as it employs no mules for the men to curse." Mr. Cathcart evidently did not hang around early railroad engineers. He claimed a running cost for fuel of a dollar per day, one third of the cost for mules. He also noted the steam engine pumped out the bilge, and the mules did not (even when your cursed them, I suppose). He goes on to cite a savings of 45 cents per ton for coal from Cumberland to Georgetown over mule power. This was one steam versus animal power revolution that fizzled out.

Horse versus Mule

The C&O Canal boats used a team of two mules, working 6 hour shifts. The second team was carried on the boat, and rested until it relieved the active team. The boat could move 24 hours per day, and sometimes were required to by the terms of the mortgage on the boat. The locks were operated around the clock. A good mule cost $125. Mules are hybrid animals that come from the union of female horses and male donkeys. Mules come in both male and female, but are almost always sterile. The size, strength, and disposition of the animals depend on their parentage. Canal traffic was limited to 4 mph, so mules were a good choice. They tend to work at their own pace, not producing as much power as horses, but able to work for longer spans of time.

Pack horses could carry some 200 pounds on the mountain paths. They don't need roads. Two hundred pounds would be, for example, 2 bushels of salt, at 84 pounds each. Each bushel could then be traded for a cow and a calf. The skin and pelt trade headed east, and raw materials and manufactured goods such as lead and iron, rifles, knives, hatchets, pots, and salt went west. Slowly.

The freight wagons on the National Road used horses. They were typically changed out at intervals, more often in the mountains. The horses provided more rapid transportation that mules. Oxen were cheaper than, but nowhere near as fast as, the horses. A rule of thumb at the time equated the pulling power of 8 horses to 12 oxen.

There are two issues in the use of draft animals (and machines, for that matter), the rate of doing work, and the endurance. Early civil engineers related everything back to horse power, from water power to steam. The definition of horse power, a measure of rate of doing work, was set by James Watt in England at 33,000 foot-pounds per minute. This was easy to measure with a horse pulling a standardized load, and timed. Watt was in the business of providing steam engines to replace horses, and used a creative financing scheme which reimbursed the user for some of the horse power replaced. This was based then on his definition of horsepower.

So, asking the question, how many horsepower can a horse produce, we have first to define what size and breed horse? A "typical" horse can produce about 1 horsepower over the course of a 10-hour day. It can produce, but not sustain, about 14 horsepower. An ox is more compliant and eats a wider range of food, but doesn't have the endurance of the horse. A mule produces less power than a horse, but has greater endurance. Sometimes it depends on what is available.

Conestoga wagons

The Conestoga wagon was a product of the Conestoga Valley of eastern Pennsylvania, and German/Swiss wagon smiths. It was designed to carry 5-8 tons, and used a team of 6 horses. The travel time from Baltimore to Cumberland was typically 7 days. A single teamster could handle the freight wagon. Besides horses, the wagons could be pulled by oxen or mules. In 1820, freight rates were roughly one dollar per 100 pounds per 100 miles, with speeds about 15 miles per day on the level. The average Conestoga wagon was 18 feet long, 11 feet high, and 4 feet in width. It cost around $65 to $100. Traditionally, it was painted with a blue underbody, and a red upper. Later freight wagons with modern features such as brakes cost about $250.

The 1755 Braddock expedition against Fort Duquesne used Pennsylvania wagons to provide logistics from the Port of Alexandria to what would become Pittsburgh. The British had brought wagons (and cannon), but found their vehicles were too heavy for effective use in the mountains west of Hagerstown. They also disparaged the American horses as being small and puny. Benjamin Franklin arranged for horses and wagons from Pennsylvania to support the military operation. Most of these were Conestoga wagons. The teamsters included a young Daniel Boone, and Daniel Morgan, later a hero of the Revolution.

The covered wagons used for the Westward expansion of the nation, the Prairie schooners, were derivatives of the Conestoga design. The Transportation Museum in Cumberland has an example of a Conestoga wagon as might have been used on the National Road. Examples can also be seen at Fort Lignoier in Pennsylvania, and the museum in Conestoga, PA.

Stage coaches

The Concord Stage Coach, built by Abbott, Downing and Company of New England, was the benchmark design. The original basic model had a twelve foot wheel base and weighed about 2000 pounds. It could be hauled by teams of 4 or 6 horses, more being needed in the mountains.

The coach itself rode on twin thorough braces formed from rawhide strips. This made a 3-inch thick leather spring for the passenger's comfort. The undercarriage was typically painted bright yellow but the coach body color was of the purchaser's choice. Typical colors were scarlet red and green. The door window was glazed but the side windows were unglazed. Canvas or leather curtains hung above each window which could be rolled down during bad weather. Typically, nine passengers and their baggage could be accommodated. The baggage allowance was 50 pounds per passenger.

Flatboats

Before the canal reached Cumberland, there was a thriving business in building single-trip Potomac river flatboats. These were constructed of white pine, and were up to 80 feet long, and 13 feet wide. They had a shallow draft, and were designed to haul up to 50 tons of coal, with four crew members. Potomac River coal traffic between 1826 and 1841 amounted to some 3,000 tons per year. These boats could only navigate the river at certain times of the year when the water level and flow permitted, and there was no practical way to get them back upstream. Thus, they also provided a ready source of lumber in Georgetown. Since June of 1839, the flat boats could enter the canal at Dam number 6, west of Hancock via a river lock and continue their journey that way. The crew made about a $10. profit per trip. Before they used portions of the C&O Canal, the river boats went all the way down the Potomac, using the skirting canals of the earlier Patowmack Canal Company at Harper's Ferry and Great Falls.

With great effort, smaller boats could be polled upstream to Cumberland in 10 days or so. Cargo was necessarily limited, and transportation costs were very high.

Canal boats

The boat captain and his family (the crew) would live aboard the boat while it was in use. After the canal was closed for the winter, the family might still live on the boat docked at Cumberland, Williamsport, or other locations. Some families also owned farmsteads. A new canal boat would cost $1-2000, and most were built in Cumberland. Used and repossessed boats were available at a lower cost. The terms of the mortgage loan usually required the operator to run the boat 24x7, until the loan was paid off. The boat builder got a construction loan to finance his cost of materials and labor. The boat was a major investment for the Captain-owner and his family.

In the early years, the canal boats were owned by the Captain, or an industrialist might own a number of boats. The coal company's also owned boats, as well as wharf facilities at Georgetown and Alexandria. Some coal company's also owned a series of sailing and steam ships to transport coal to New England. The big companies, then, were involved not only in the production but the transportation of the raw material. During 1852 the Cumberland Coal & Iron Company invested in canal boats starting with an initial purchase of 31 boats, then increased their fleet to 68 boats in 1856. Borden Mining Company initially bought 8 boats but later sold them in 1852. The other coal companies never had more than 4 boats in this time period. A peak of some 500 boats were registered on the C&O Canal. After the Canal Towage Company took over ownership in 1902, most boats were company property. This marked the end of most independent owner-operators on the canal, and boat Captains became employees.

Intermodal wins

In discussing the 19th century canal and railroad interactions, we use the word intermodal to describe the movement of a bulk commodity, such as coal, between rail and canal modes of transportation. At Cumberland, coal reached the canal docks via train, and was transferred to the boats. Coal cars could also go from the mine site directly to the user, or to ships for export. Trans-loading is time consuming and expensive. Canal boats were emptied by hand-shoveling in the early days, later by steam or water powered derricks. Initially, loading from coal car was done by dumping the coal, then hand shoveling into the canal boat. This was all very labor intensive. Some goods, such as flour or cement, moved in standardized containers, barrels.

Today's term intermodal generally refers to standardized cargo that can travel by road, rail, or water. What has made the efficiencies possible is the development of the standardized shipping container. These are large, strong, standard sized boxes that allow for material to be shipped, but only handled when the container is loaded or unloaded. The efficiency comes in using handling and transportation equipment that is built to one standard. Today, even liquids and bulk material travel in standard containers that are reusable. The containers are covered by International Standards, allowing for efficient transport anywhere in the world. The notable exception is by air. Barrels, made to a standard size, were the precursors of today's shipping containers. The rise of the cooperage (barrel making) trade flourished when goods needed to be shipped. Pallets, developed during World War-II to address the standardization of cargo, were a step towards today's containers. Another intermediate step was the carrying of road trailers-on-flat-cars (TOFC) by the railways, allowing for trailers and not their contents to be handled. En route, the road wheels are not needed, so the standardized container developed out of the truck body. The Department of Defense led the effort in the 1950's.

In today's railroad freight shipping, there are basically two types of trains. The unit train moves the same cargo from a source to a destination, for example, coal or automobiles. Special rail cars are used. Intermodal containers move on their own special cars, which carry multiple containers stacked two high. Some rail lines cannot accommodate double-stack containers, due to bridge or tunnel restrictions. Easing this restriction can provide a theoretical doubling of capacity. CSX's National Gateway Project, kicked off in 2008, aims to remove restrictions to double-stacked freight shipments across their system. It is a private-public partnership. A single restriction can prevent an entire line from being used with two layers of containers. One major problematic point is the 19th century Howard Street tunnel in Baltimore. Intermodal terminals, like Norfolk-Sothern's Virginia Inland Port at Front Royal along Interstate-81 allow for distributing a train-load of containers to fleets of trucks for final delivery. In the other direction, truck-loads of containers make up one train that goes directly to the Port of Norfolk, for loading on ocean-going vessels.

Traffic Analysis

<u>What cargo moved on the canal?</u>

Way bill records remain from the Canal Company regarding tariffs collected on cargo. These give us an insight on the types of freight that moved upstream and downstream. Just as with any transportation system, it is more efficient to move cargo in both directions. But, you can't haul wheat in a boat that just hauled coal in bulk. You can't haul dry cement in a leaky boat. We have to distinguish between water-insensitive bulk cargo like limestone, ore, or coal, and water-sensitive cargo such as grain and cement. Most of this type would have been packaged in barrels. Thus, we would expect to find thriving cooperage works at both end of the canal.

Generally, upstream cargo was less than 10% of the downstream cargo on the canal. What moved downstream was raw materials. What moved upstream was finished goods. Thus, the manufacturing centers and centers of distribution were downstream (Alexandria, Georgetown), and the upstream locations were centers of production of raw materials (coal, agricultural products). But, the Industrial Revolution moved westward as well. It was not unnoticed that it was, in the long run, cheaper to move the centers of production closer to the source of materials (and energy). This was because it was more economical to ship manufactured goods than raw materials. This worked as long as cargo flowed downstream to distribution centers along the Atlantic seacoast. But, the country was growing. More goods were required west of the Alleghenies. More agricultural products came from west of the Ohio River. A transportation system that economically crossed the mountains was needed.

Maybe the French had gotten it right. They had colonies in Canada and at New Orleans, connected by north-south flowing rivers. The English settled the eastern seaboard, and pushed west, with mountains in the way. In terms of internal transportation, it was easier to move men and materials north-south, than east-west. The mountains would be conquered by technology - that of steam. The focus of trade changed from north-south to east-west, enabled by the new transportation alternatives.

Tolls were set by the Canal Company in the period 1830-1850 at the same rate that the earlier Patwomack Company charged for use of their skirting canal around Great Falls. The cargo items mentioned included domestic spirits, tobacco, linseed oil, wheat, peas, beans, flax seed, corn, flour, pork and beef.

Bulk items were charged by the (long) ton. These included hemp, flax, potash, and ores of iron, copper, and lead. Coal was priced by a British volume measure, the *chaldron*, about 3,100 pounds of coal. Timber and planking were charged by the board-foot. Most miscellaneous cargo was charged by the hundred-weight. Tobacco was charged by the hogshead; grain, fruit, salt and such by the bushel. Fish were charged by the count. Other items for which a tariff was established included grindstones, brick, tile, and roofing slate, cast or pig iron, and wood bark. The tariffs give us an idea of what materials were in demand, and were worth shipping. If there is an established tariff, then that product was important enough to ship.

In 1841, we see fence rails added to the list, as well as wooden shingles. In 1843, furniture was given its own tariff, as was worked metal. Carriages, wagons, and plows were added to the list, as was window glass. Oysters and ice appeared. New markets were being developed, and new industries were established. There was a tariff for dry goods and sundries.

It took the canal 22 years to get from Georgetown to Cumberland. During this time, the canal mostly moved agricultural products, and building materials for its own use. Prior to 1850 agricultural products flowed into the canal through the river locks at Edwards Ferry and Sharpsburg from Loudoun County, VA. Loudoun County, VA and Montgomery County, MD farmers brought their products to the granaries at Edwards Ferry, Whites Ferry, Seneca and Monocacy where the grains were put in separate bins for subsequent shipment to Georgetown by canal boat. Shelled corn from several farms was co-mingled in the bin until there was enough to make a load. The same was done for wheat, rye, and oats. After the grain was sold in Georgetown, the farmers were paid for their share. As the canal pushed further west, more lumber as downstream cargo was seen.

In 1843, the B&O began hauling Cumberland coal to Dam #6 for trans-loading onto canal boats. This would continue until 1850, when canal boats could finally reach Cumberland. In the 1850's coal traffic grew to dominant the downstream trade, growing to 80% of the total goods carried. In 1860, the upstream traffic was 3% of the downstream traffic. Partially, this was because coal dominated, but also fewer manufactured goods were needed in the west, as the industries there developed and matured. The west became self-sufficient in manufactured goods and food; materials were moved when economical to do so.

Downstream freight

There were 3 choices to move heavy freight from Cumberland before the Civil War:

1) Freight wagons on the Turnpike east, and on the National Road west. This took 2-3 days, with each wagon carrying 3-5 tons. The traffic slowed but didn't stop in winter. The National Road was of Federal construction. Tolls were charged by the State for road maintenance. The National Road ran westward from Cumberland. Earlier, in the 1804 time frame, the State incorporated a company to build a road from Baltimore to Frederick. The company would be authorized to collect tolls for its maintenance. The Baltimore to Frederick Turnpike cost $8,000 per mile to construct. It was extended west to Boonsborough, but still had 74 1/2 miles to go to Cumberland. The State, faced with war debts from the War of 1812, needed to get creative about financing.

So, it came up with a clever scheme where the renewal of a Bank's State charter was tied to its participation in a road building company. The City Bank of Baltimore, the Hagerstown Bank, the Conococheague Bank, and the Cumberland Bank of Allegany formed the Cumberland Turnpike Company. The right-of-way was sometimes called "the Bank Road."

2) The B&O Railroad - There were several trains a day to and between Cumberland and Baltimore. The journey took some 10 hours, and the capacity of a freight car was 10 tons. Cargo could be handled in any weather. The B&O was privately owned, but with bonds held by the State of Maryland, and the City of Baltimore. Passenger service was provided on the line. Originally single-tracked, the B&O hastened to put more parallel track in place to expedite train movements. Some of this was purchased in Mount Savage. The B&O's rolling stock and motive power was built in Baltimore.

The B&O initially focused on the passenger and flour trade. Coal was not really considered. When coal was needed it could more easily go on the canal. When the Maryland Mining Company asked the B&O to quote a cost to transport 100,000 tons per year, the B&O began to realize there might be a business model for coal. Maryland Mining, with its New York interests, already knew that. Coal in Eckhart was a transportation problem; coal in New York was a profit. The Maryland and New York Iron & Coal Company asked the B&O for the cost to ship 1,000 tons per day. These would go the Dam 6, and then trans-loaded to the canal.

Coal was carried in open top, bottom-dumping iron pot hoppers. Unit 23001 at the B&O Museum in Baltimore is an example of a later 8-wheel Winans design. These have a tare weight of 16,350 pounds, and a capacity of 40,000 pounds. Other goods traveled in iron box cars. Unit 17001, at the B&O Museum in Baltimore is an example of one of these. It has a tare weight of 20,000 pounds, and a capacity ranging to 15 tons. Dry bulk goods such as flour and cement traveled in barrels. A modern dry barrel holds 7,056 cubic inches or 3.28 bushels. The weigh varies depending on the density of the contents.

A bushel is 8 gallons of dry measure. For corn, a bushel weighs 56 pounds, so a barrel would hold just over 183 pounds. Wheat weighs more, at 60 pounds per bushel, or close to 197 pounds per barrel. Call it 200 pounds, with the empty weight of the barrel. The box cars, volume permitting, would hold up to 150 barrels. Cement, depending on the fineness of the grind, can weigh 120 pounds per bushel, or close to 400 pounds per barrel. A box car would hold 75 barrels of cement, limited by weight.

3) By canal boat, the journey took 5 days with 120 tons of cargo. A typical canal boat was 14 1/2 feet wide and 92 feet long. The rudder swung up so the boat would fit in the lock. A boat would draw 4-5 feet of water. The canal shut down in winter due to freezing. The company was private, but with bonds held by the States and individuals. The canal had the lowest transportation rate, and was ideal for time-insensitive bulk freight such as coal. However, when the demand for coal peaked in the winter, the canal was not usable. In addition, the spring floods damaged the canal, and extensive repairs sometimes shut the system down for weeks or months. The canal was highly dependent on the fickleness of Nature.

There are detailed records of the canal trade from 1831-1850, and 1851-1878, based on tariffs charged. Agricultural products moved downstream. These included tobacco, wheat, flour, rye, bran, corn and corn meal, flax, hemp, and whiskey. We also see pig iron, leather, cement, bacon and lard, wood and stone. Ascending cargo included fish, salt and plaster. Going upstream in 1845 were bricks, coal, oysters, apples, salted and fresh fish, whiskey and potatoes. Coal began to flow downstream around 1846 from Dam #6, and by 1850 from Cumberland. Pig iron and iron ore were shipped as well. Thousands of kegs of nails are mentioned. In the years 1851-1878, there are 38 categories of freight. Tobacco had become insignificant, but empty barrels, manufactured furniture, wrought iron, beer, and slat wood picked up the slack. Coke shipments dropped off in 1866, but then coal totally dominated the downstream traffic.

The B&O Railroad showed similar cargo going down to Baltimore, again with coal coming to dominant the shipments. It didn't even report cargo going up to Cumberland; almost all of this traffic was empty coal hoppers.

Flour shipments

For quite a while, flour was a major downstream freight. It had been the driving force for construction of the canal and railroad system in the first place - to tap into the Shenandoah Valley and the lands west of the Ohio, where agricultural products were plentiful. This, in turn, was Washington's vision of a transportation mode, a "method of internal communication" between the coast, and beyond the mountains to the Ohio River. Eastern mills needed product. Mills had to be built near a power source, and this was flowing water. Mills in Georgetown used water from the canal (metered, and charged for) to grind grain from the west. Between 1840 and 1850, flour was the major trade along the eastern seaboard and for export, originating at Georgetown. Flour went to the West Indies and Brazil. From 1848 to 1853, over 200,000 barrels of flour per year were shipped on the canal. In 1844, a cotton mill at Georgetown was converted into a flour mill to meet increasing demand. In 1856, Georgetown had five flour mills on the canal, as well as the works of the Pioneer Cotton Company.

In 1884, southwestern Pennsylvania counties began using the canal to ship their agricultural products. In 1832, the Baltimore & Ohio Railroad shipped almost 150,000 barrels of flour. This rose to almost 775,000 barrels in 1862. The first collision between the canal and the railroad over cargo occurred at Harpers Ferry, the choke point. It was the rivalry over the wheat and flour trade from the Shenandoah Valley in the 1840's.

Mills in Alexandria were the growth industry from the time of George Washington. If he hadn't been distracted by the Revolution and the Presidency, he may have become the outstanding entrepreneur of his age. He had the far-sightedness to see the value of a transportation system east-west across the mountains. Washington owned a mill site in Alexandria, and a brick mill was built there by William Sheppard. A lot of Alexandria-ground flour was exported through the port at the end of the 18th century; around 50,000 barrels of flour were exported per year. This was also good for the cooperage business. Early in the 1800's, Alexandria shipped 1,000,000 barrels of wheat flour to Europe and the West Indies. By 1810, Alexandria was the hub of an industry of flour production, with over 100 mills, and the transportation infrastructure of roads, canals, the Potomac River, and the Chesapeake Bay.

There was an early canal-railroad price war over flour shipments. Flour was important to the Port of Baltimore. This persisted until 1848, when coal took over. One-half million tons of flour went east in the first 20 years of the trade. The transportation cost was 4 cents per ton.

Coal

Philip E. Thomas, first president of the B&O railroad, noted in 1836, " It had not been the intention of our Company to make any provision for the transportation of coal, as that article can more appropriately be conveyed on the canal."

The Federal Government was a major consumer of coal, and coal demand exploded during the Civil War, as more new-fangled steam ships were being built. The Union iron clad ship Monitor, after her battle at Hampton Roads, came back to the Washington Navy Yard for refitting and repair, and was restocked with George's Creek coal.

Major manufacturing sites in the industrial enclave of Georgetown also required coal. The Foxall Foundry used water power from Rock Creek to operate its cannon boring machinery, but needed coal in large quantities to melt bronze and iron for cannons and shot. The foundry is described as a major defense contractor and part of the country's Military-Industrial Complex. From 1800 to 1809, it was the sole supplier of big guns to the Government, along with shot and gun carriages. It became a target of the British Invasion fleet in the War of 1812, but, although the White House was burned, the foundry escaped damage. It was built, owned, and operated by Henry Foxall, who was an iron founder from England. He moved to Georgetown at the personal request of Thomas Jefferson, when the Seat of Government moved to the new city of Washington from Philadelphia.

George Mason purchased the facility in 1815; it was known as the Columbia Foundry after 1816. The foundry process started with pig iron, and it is possible, but not documented, that iron from the furnace at Lonaconing in Western Maryland was used. The facility did have its own blast furnace, which would have needed large quantities of iron ore, coked coal, and limestone. For the molding process, clay and sand were required. The cannon were cast solid, then bored, as was the practice of the day. This produced more accurate cannon than the hollow-casting method. Other machinery at the facility included screw cutting machines, and lathes. The boring machines were horizontal and used water power. One was capable of handling nine cannon at once. The facility was sold after Mason's death in 1849, and stopped production.

DuVall's Foundry in Georgetown on 30th Street operated from around 1856 to the Civil War. It was started in response to the House Committee on Military Affairs' recommendation for a National Foundry in Georgetown. It used coal from Cumberland and pig iron from Antietam. William T. Duvall operated the facility for the Government. The building still stands, at Lock 3, in mixed use as shops and offices.

What happened to the coal from Cumberland?

When the B&O was severed at Harper's Ferry by Stonewall Jackson, the canal from Cumberland became the coal lifeline to the Nation's capital. Coal was essential for the U. S. Navy's steamship fleet, and George's Creek coal was preferred. At the river lock, loaded canal boats could enter the Potomac, and be towed by steam tug to the Washington Navy Yard, a major shipyard and ordnance facility in Southeast DC. Established in 1799, the Washington Navy Yard became the navy's largest shipbuilding and ship fitting facility, with 22 vessels constructed there. It hosted the Naval Gun Factory. The USS Constitution and the USS Monitor were both refitted there. It served a vital role in the defense of Washington during the Civil War. Around 1845, the Atkinson and Templeman Company contracted with the federal government for the delivery of 40,000 tons of western Maryland coal to the District.

Coal was trans-shipped at Williamsport from the canal to the Western Maryland Railroad for delivery to Pennsylvania and Baltimore, and to wagons for local use. Coal went to Harper's Ferry for use by the Federal Arsenal and the industries on the manufacturing island of Virginius.

Canal boats went out into the Potomac River via the tide lock to the Washington Navy Yard, the Naval Coaling Station Indian Head, and to the Port of Alexandria for loading on ships with destinations of New York, Boston, and New England's textile mills. The ships went down the Potomac to the Chesapeake Bay, then south to the Atlantic Ocean, or north to the Delaware & Maryland Canal, then to the Atlantic and points north. Export coal went to the British West Indies and to a British Navy coaling port on the north coast of South America.

Canal boats went via aqueduct across the Potomac to the Alexandria Canal and the Port of Alexandria for export. In 1852, 287 sailing vessels, 1 steamer, and 23 barges loaded with Cumberland coal left the port. The Alexandria Canal owned the aqueduct.

Coal went via the Baltimore & Ohio Railroad to the Port of Baltimore. Besides local use, in 1852, Cumberland coal left the Port on 978 sailing vessels, 5 steamships, and 130 barges. George's Creek coal was much in favor by the big transatlantic passenger lines. Cumberland coal was preferred by the owners of the steamship Great Eastern. Edward Cunard was one of the founders of the Lonaconing Ocean Coal Mining and Transportation Company, later the Ocean Steam Coal Company. Ocean, a small cluster of houses along the George's Creek, was the deep mine site.

Enabled Industries

The existence of transportation options enabled industries that would not have been economically feasible before. The difficulty of getting heavy supplies inland was aptly demonstrated by the logistics of the Braddock Military Expedition of 1755. Receiving his military supplies from England at the Port of Alexandria, his challenge was to get them to Fort Duquesne, modern day Pittsburgh. This resulted in the construction of the Braddock Road. The transportation difficulties were seen first-hand by Washington.

Transportation methods allowed farmers to sell their products to markets farther away. Conversely, the larger cities on the seacoast could obtain their food and building materials from other than local sources. Bulk items such as building stone, timber, and coal could be economically shipped. Items available only along the seacoast such as oysters and crabs could be made available in western markets. Inland, there was a demand for salt.

Along with this, the transportation systems created their own maintenance and supply requirements. The canal needed mules, harness and tack, boats, supplies for the boatmen and their families, lumber and stone for maintenance. The railroads needed coal, locomotives and cars, rail, switches, iron, lumber, and many other supplies. Both needed a large number of barrels, to ship cargo such as flour, cement, and gypsum. It was a synergy of supply and demand. The transportation methods enabled and demanded new industries, which in turn required more raw materials, and better transportation methods.

The transportation revolution also brought about improvements in communication. With good roads, and a railroad, correspondence traveled faster. The mail didn't require fast horses any more, but rode on canal packet boats, and trains. With the advent of the telegraph, communications became much faster and reliable. The railroads embraced the telegraph early on, building the lines along their right-of-way. This allowed for more efficient dispatching and operations, and better response to accidents and problems on the line. The canal installed a telephone system in 1879. According to Unrau, "The telephone system, which was the longest single circuit then in existence, enabled the canal company to reduce operating costs by providing fast communication of information relative to breaches and canal traffic problems. Such information had been carried formerly on foot or by horseback or mail." Here are some examples of industries taking advantage of transportation options to enable or grow their business."

Marble for the Nation's Capital traveled via rail from a quarry near Cockeysville, Maryland, on the B&O. The Seneca stone cutting mill at MP 22.8 on the canal provided the sandstone for the Smithsonian Castle, and numerous other buildings in the District of Columbia.. The mill also used water from the canal as a power source. Red Seneca sandstone is the basis for the then-popular brownstone architecture. In fact, the locks of the Patowmack Canal at Great Falls are constructed of this stone, and it was used in the structure of the Washington Monument. Bricks also made the downstream journey in vast quantities.

Cement works along the canal as far away as Cumberland shipped their product in barrels for lock construction. Rough stone was needed for construction of bridges, aqueducts. wharfs, dams, and other structures. Pig and scrap iron made the journey, as did timber from the western part of the states.

Few large trees survived near major cities due to the demand for firewood. After coal was introduced as a fuel, the demand for firewood vastly decreased. The demand for wood for construction grew.

On March 12, 1838, The Maryland and New York Iron & Coal Company was incorporated in Maryland. This was in anticipation of the arrival of the canal and the railroad at Cumberland. The State charter allowed them to build or acquire railroads, as long as they did not interfere with the B&O, or the C&O Canal. They were also required to erect an Iron Works at Mount Savage, and produce

1000 tons of pig, cast, or bar iron in one year. It is important to note that when the legislature granted the rights to build a railroad, they included the right to condemn and acquire private land if it were needed to build the line.

The charter said, "…it shall not be lawful for the said Maryland and New York Iron and Coal Company to occupy or use any portion of the lands that may be necessary for the accommodation of the canal and works of the Chesapeake and Ohio Canal Company, or for the main route of the Baltimore and Ohio Rail Road."

The blowing engines for the blast furnaces at Mount Savage came from the West Point Foundry in New York in 1845, as had the ones for the furnace at Lonaconing. They were sized for making 400 tons of iron per week. Then engines were of the condensing type (recycling water), with a 56-inch diameter cylinder and a 10-foot stroke. They made 15 revolutions per minute, using steam at 60 pounds per square inch and generating 80 horsepower. The associated boilers were 60 inches in diameter and 24 feet long. The grates spanned a total of 198 square feet. The blast cylinders were massive, being 126 inches in diameter with a 10 foot stroke. They operated at 15 revolutions per minute, and supplied air at 4-5 pounds per square inch pressure. One engine was used for the blast furnaces, and the other for the rolling mill. At the time, they were the largest cast cylinders in the world. The shipment went to Georgetown via sailing ship, then up the canal as far as it could, which was Williamsport. The canal had frozen, and the machinery shipment had to proceed to Mount Savage by horse and wagon.

When Mount Savage had installed the rolling mill and built the Mount Savage Rail Road to Cumberland, its product was supplied to other railroads. It was sold to the B&O Railroad, which up to that time had been dependent on imported British rail. One thousand tons of rail, at $59 per ton, went to a railroad at Fall River, Massachusetts. An additional customer included the Hampshire Coal & Iron Company for their tram road near

Piedmont, WV. The Utica & Schenectady and the Hudson River Railroad in New York, and the Erie and the Reading in Pennsylvania were also customers. The Utica and Schenectady Railroad ordered 1,000 tons of rail. There was a display of Mount Savage rail at the Mechanics Fair in Baltimore in November 1850. E. Pratt & Brother of Baltimore were the agents for the Company.

The earliest fuel for the B&O locomotives was wood. The B&O used 3,400 cords of wood in 1841. Trees were getting scarce. The transition to coal was slow, and anthracite coal from Pennsylvania was initially popular. By 1838, Latrobe had shown that the company's locomotives could successfully burn the Cumberland coal. Different fuels, hard coal, soft coal, wood, required different grate and furnace designs.

The Moore & Waters Company shipped guano up the Canal for use as agricultural fertilizer. It mostly came from islands near Peru. Caribbean guano went to the Port of Baltimore. Guano is a waste product of seabirds. A good source of nitrates, it was in large demand for use by the military. The Guano Islands Act (11 Stat. 119, enacted 18 August 1856, codified at 48 U.S.C. ch.8 §§ 1411-1419) was federal legislation passed by the U.S. Congress that enabled citizens of the U.S. to take possession of islands containing large guano deposits. The islands could be located anywhere, so long as they were not occupied and not within the jurisdiction of other governments. It also empowered the President of the United States to use the military to protect such interests, as guano was considered a strategic material.

Melchior J. Miller was the native-born son of a German immigrant and farmer who arrived in the United States in the early 1830s. In 1875, Miller purchased a farm along a tributary of South Branch Bear Creek, just southeast of Accident, MD. He also bought out the equipment of a small distillery and moved it to his farm. He hired professionals to operate the business. His three sons, William, John, and Charles, learned the trade from these experts, eventually replacing them. In 1902 Melkey sold the distillery to his sons.

William continued as a distiller, while John and Charles established a wholesale and retail whiskey business in nearby Westernport. Whiskey could be sent downstream by boat to Cumberland where it could either continue down the Chesapeake and Ohio Canal to Washington, be carted by wagon over the National Road to Baltimore, or later go by railroad. The company changed its name to M.J. Millers Sons Distillery. The sons built "Melky Miller's Maryland Rye Whiskey" into a highly respected local and regional brand. Production was relatively small at 29 bushels of grain processed daily according to Federal records. Unfortunately, the passage of the Volstead Act (Prohibition) in 1919 brought the family business to a close. Mostly,

West of Cumberland

The canal never was built west of Cumberland, but the optimal paths to the Ohio River had been determined. The Patowmack Company did a survey for a waterway along the southern route in 1784. As it turns out, the B&O railroad used two of the possible canal paths on its way to the Ohio. This is not surprising, as there are few good options through the steep grades of the Alleghenies, and no complete water-level route. The Congress had in 1824 authorized the President "to cause the necessary surveys, plans, and estimates to be made of the routes of such roads and canals as he may deem of national importance in a commercial or military point of view, or necessary for transportation of the public mail,..." this section discusses the options of continuing the canal westward and why it was planned but never built. Again we see the competition for the right-of-way being contested by different modes of transportation.

The survey of potential canal paths west by the Corps of Engineers in 1876 found the prior surveys of Col. Thomas S. Sedgwick valid. He, in turn, had used the prior work of Captain William G. McNeill, Topological Engineer, under the direction of the Board of Engineers for Internal Improvement, in 1824. This extended a series of surveys by the Patowmack Company. These, in turn, went back to the surveying work of Washington for Lord Fairfax, and the detailed knowledge of the terrain by the agents of the Ohio Company (Gist, Cresap, and others) The new surveys included considerations of the sources of water, the cost of construction, the time required, and the relative merits of the routes. Sedgewick relied heavily on "lessons learned" of the Europeans, particularity the French, canal system (Graeff, Construction des Canaux et des Chemins de Fer, Paris, 1861). A comprehensive study was submitted by Brigadier General A. A. Humphreys, Chief of Engineers, to the Secretary of War, and submitted to the 44th Congress. The building of the western canal extension continued to be of interest during World War I and as part of the New Deal Public Works Projects. But the harsh reality of the costs always prevailed.

George Washington was a staunch supporter of trans-Appalachian routes for commerce. Before the Revolution, he saw the potential opportunity of opening up the lands for commerce. He explored ways of connecting the Potomac and the Ohio, even if it involved portages. His estate, Mount Vernon, was on the Potomac, and convenient to the Port of Alexandria. Washington had acquired lands in "the West"," and received more in Western Pennsylvania as a result of his service in the Revolution. After he resigned his commission in the Army, he turned back to farming, and his interest in his western lands. He set off again in 1784 to search for the elusive passage to the West, and to collect back rents

Washington discussed the Patowmack Company with his peers, a distinguished group of wealthy landholders in Virginia. Then, his country called upon him again for service in 1789, as the first President. This he reluctantly accepted, as this took his attentions away from the West. He was succeeded by Thomas Johnson as Patowmack Company president. He continued to worry that without adequate means of commerce and communication, the country west of the Appalachians might split off and become another nation, influenced by the British, the French, or the Spanish.

The Patowmack Canal Company didn't get much done for the first twenty-five years or so. The company's concept was to use the Potomac for slackwater navigation, with skirting canals around the difficult parts, which included Great Falls and the Harper's Ferry area. These were actually built, although at much more effort and expense than anticipated or planned for. Even these improvements, along with the little and Great Falls canals, enabled a lot of merchandise to come down to tidewater.

Later, the charter of the Potomac Company would be transferred to the C&O Canal Company. The new concept was for a parallel canal, not utilizing the Potomac for anything but a water source.

This was painfully shown to be not a great idea either. The Potomac frequently had either too much or not enough water. In its 36 years, the Patowmack Canal system cost $729,000 to build and maintain.

Washington's vision was a good one, not totally supported by the technology of the day. Washington had seen a demonstration of an early concept for Rumsey's steamboat. It is not clear that he had ever seen or heard of a railroad. He was open to new technologies, but the required ones would mature after he was gone.

The two options to head west from Cumberland involved different routes which we'll call the north and the south. The northern route went up Will's Creek from its junction with the Potomac, through the Narrows, and into Pennsylvania. Then it headed west up and through the mountains, to the Castleman and Youghiogheny Rivers, then to the Monongahela River and Pittsburgh. The southern route is up the Savage River to Deep Creek to the Youghiogheny River and its junction with the Castleman River. From Georgetown to Pittsburgh, the elevation gain was 3,837 feet. This was more than what had been considered feasible up to that time.

The survey team also went to the headwaters of the Potomac at the Fairfax Stone, in order to explore an option using the Black Water fork of the Cheat River. Unfortunately, this expedition had to turn back due to excessive snowfall, not uncommon in that area.

Canal Extension, option 1

This preferred alternative followed the course of Will's Creek upstream out of Cumberland and through the Narrows, then up to Pennsylvania. Turn west, and then go up the mountain. The design guidelines were to the same parameters and dimensions as the Chesapeake & Ohio canal used in construction to Cumberland.

The Narrows passage dominates the west end of the Cumberland skyline. This valley, the gateway to the west, is a natural geologic feature carved by Will's Creek. The cliffs of Lover's Leap rise some 800 feet above Will's Creek on Will's Mountain to the North. Haystack Mountain is to the south. It has taken Will's Creek some 150 million years to wear down the rock of the Allegany mountains, up-thrust from the ancient seabed, into the two separate mountains.

The strata revealed in the rocks is of the Juniata Formation, 530 feet thick and predominately red, dating from the Ordovician period, some 435-460 million years ago. Overlaying this is some 380 feet of the younger Tuscarora sandstone, from the Silurina period, 425-435 million years old. The strata is up-thrust at the ends, most noticeably at the west end of Will's Mountain at Locust Grove. From Haystack Mountain to Will's Mountain, the valley is about 1/2 mile wide across at the top of the cliffs.

Passing through the Narrows now are Will's Creek and the National Road (Route 40), as well as several rail lines. At some time, all of these rail lines went through the Narrows: Western Maryland, Pennsylvania, Cumberland & Pennsylvania, Baltimore & Ohio, CSX, Chessie, George's Creek & Cumberland, Eckhart Branch, Mt. Savage Rail Road, and the Cumberland & Westernport Electric Railway trolley line. This choke point is the only option besides going over Haystack Mountain. That option is too steep for the rail lines or the canal. How would the canal have proceeded through the Narrows? Currently, and for the past 100 years, there has been two sets of rail lines, the National Road, and Will's Creek crammed into the passage. The canal builders chose to use the creek bed. This would be a good option at all times except the spring floods. The Eckhart Branch rail bridge would have had to be modified. The plan was to enclose the Creek in masonry walls and place the canal in the center. I suspect this would have lasted until the next flood.

Essentially, the canal would have followed the path of the Pittsburg (sic) and Connellsville Railroad (later, B&O). This makes sense;

the railroad and the canal faced similar problems with grade and access. However, the railroads did not need the assured sources of water at the top of the grade. The 1875 Survey started with a request from the Government Engineer, Major Merrill, to Benjamin Latrobe, Chief Engineer of the Baltimore & Ohio Railroad, for the information acquired in the surveys for the rail line. These were politely supplied. Why not? The rail line had already been built. As the canal engineers found out, it was built in the best path. The canal was too late. The Board of Internal Improvements Report of 1826 showed the best route was Will's Creek to the Castleman River to the Youghiogheny River. This path was shorter than the other options, and required a lower summit tunnel. However, this optimum route had rail tracks on it in 1875 - the Pittsburg and Connellsville Railroad. The tunnel was the Baltimore & Ohio's Sand Patch tunnel. Latrobe related that the tunnel had cost $420,000 to construct its 4,800 foot length. This gave the canal engineers a reality check on their own estimates.

"I am decidedly of the opinion that the summit-tunnel should be worked by steam."

If they eliminated the tow path, the tunnel would be less expensive, based on the French experience. Either tug boats or stationary engines and tow cables would be used to move the boats. As the railroads had found out, long tunnels worked by steam engines tended to fill up with smoke rapidly. Five tug boats were specified for use at the canal tunnel.

This option was considered at the canal tunnel at Paw Paw, WV, but not implemented This approach would still not have allowed for two boats side-by-side, but would have allowed more water in the channel, and the boats to ride higher, thus reducing their drag. Stationary steam engines were to be used to pull the boats through. The European experience with tunnels was to group the boats into "trains". One European tunnel considered was 13 miles long (the Paw Paw tunnel is 5/8 mile).

A major worry of the canal builders was an adequate water supply to replenish losses due to evaporation and leakage. A certain amount of water is lost downstream every time a load is locked through. Evaporative losses were well known. Loss due to leakage could be countered by cementing the path of the canal, an expensive proposition. The calculated figure for water needed was 42 cubic feet of water per mile, per minute. This was a bit of a problem on the eastern slope, but the engineers thought the Casselman River could handle the supply on the western slope. They assumed 150 lockage transits per day, requiring 87.5 cubic feet of water per running foot, per 24 hours. There were to be two reservoirs on the Castleman River above Salisbury, PA.

There was a trade-off at the top of the mountain, between continuing the lift system (locks or inclined planes) to the summit, or using a summit tunnel to lower the required lift. There were cost and time trade-offs. A summit tunnel was the chosen option. One can get canal boats up and down a mountain by several means. As long as water is available for the operation of the locks, that scheme will work. Inclined planes are another method, used at Georgetown, and on the Allegany Portage Railroad in Pennsylvania. The Georgetown Plane used a caisson, a water-filled container to hold the canal boat. This was essentially a portable lock. There are pros and cons to each approach.

Sandpatch grade is located on the B&O rail line to Pittsburgh, west of Cumberland. It features steep grades and a long railroad tunnel. At Mt. Savage Junction, J Tower once guarded the interchange with the C&P, and the Pittsburg and Connellsville (P&C). At Hyndman, PA, Q tower marked the beginning of the 1.88% grade.

The railroad tunnel at Sandpatch was begun in 1854, and opened in 1871. It was single-tracked, and 4,777 feet long. In 1911, a new tunnel was begun. Only 4,475 feet long, but double-tracked, it was opened in 1913. The canal tunnel would have been deeper and longer. Where the tunnel was built also affected its length. Closer to the top of the mountain, the tunnel would have been shorter, but more locks or inclined planes would have been needed to reach it.

The summit tunnel, based on the survey of 1826, was to be located at an elevation of 1,972 feet. This was modified later to be some 28 feet lower. This would have had the effect of requiring a shorter feeder tunnel from the reservoir. The proposed tunnel was larger (28 feet high, 46 feet wide) than any then in use. It was designed to allow 2-way traffic, which was thought to be essential, based on the Paw Paw experience.

From Cumberland to the summit tunnel there would have been 17 inclined planes, for a total lift of 1,185 feet. This eliminated the need for 148 additional locks. The western slope was more gradual, with six planes, and fifty-six locks. The plan was to go up the Youghiogheny River as far as Connellsville, then head up river to West Newton, PA. This slackwater option was appealing. Some dams and locks would be required. The eastern slope would have also required 5 aqueducts, with 6 more on the western slope.

The shipment of coal, coke, and iron ore from Connellsville to Pittsburgh was also mentioned for the westernmost portion of the canal. The accommodation of the type of river barge used on the Ohio (125 feet long, 25-30 feet wide) to Connellsville was also factored in. The question of importance was, what would be the motive power for the large barges? It was not clear at the time whether steam power would ever replace animal power.

The summit reservoir, as proposed in 1826 (the Pleucher Reservoir), would be built along with a smaller reservoir closer to the canal. These would have a combined capacity of some 250 million cubic feet of water.

The estimated cost of the canal extension from Cumberland to Pittsburgh by the Will's Creek and Youghiogheny River Route was estimated to be $25 million dollars. It had cost $11 million to get the canal to Cumberland from Georgetown.

Canal extension, option 2

The second option for getting the canal to Pittsburgh involved following the Potomac to the confluence of the Savage River, then up the Savage River to the Castleman river. From Cumberland to Westernport, and a mile or so beyond would have been easy. The route is water-level, along the Potomac River. Shortly thereafter, Backbone Mountain is in the way. An early design involved a long tunnel, with a feeder lake in Garrett County. The tunnel would have been longer than any attempted yet in the world. Curiously, the lake would later be built as a source of hydroelectric power, and recreational boating. It is called Deep Creek Lake, an artificial lake of 3,900 acres extent. The proposed summit reservoirs for the canal were to have a capacity of over 252 million cubic feet, and a surface area of some 200 acres.

The report mentions that the North Branch Route, was inferior to the Will's Creek one by having a higher summit level and a longer required tunnel. The local line extension from Cumberland to at least the confluence of George's Creek at Westernport would be important for the coal trade. This would save 28 miles of rail transportation, but rail transportation was established by then. The section from George's Creek to the Savage River was going to be problematical, due to existing bridges, mills, and the town of Piedmont. Senator Davis would most likely have supported the construction. It fit well with his West Virginia Central and Pittsburg Railroad, his Piedmont & Cumberland Railroad, Cumberland and Piedmont Railroad, Potomac & Piedmont Coal & Railroad Company, and the Bloomington & Fairfax Railroad. The canal extension would have added 30.5 miles to the canal from Cumberland. The drop in elevation is 335.3 feet, required 43 locks. The Savage River would have been used as a feeder water supply.

This option would have a dam across the Potomac some 600 feet below the mouth of the Savage River. From the dam it was about 1/2 mile to "the Honorable H. G. Davis' sawmills." Boats would pass into the Potomac at a river lock upstream of Piedmont. The

C&P railroad bridge crossing from Westernport to Piedmont would need to be raised or relocated. An alternative taking the canal through the center of Piedmont was considered and discarded, because of potential conflicts with the Baltimore & Ohio Railroad. The Luke Paper Mill was built in 1888. It required large amounts of pulp wood. It occupied land that could have been used by the canal.

Another sub-option would have been to terminate the canal before Piedmont, and use an extension of the C&P Railroad to reach it. Lock 19 would have been the Keyser lock. Keyser, WV, once known as New Creek, was the site of major B&O yard facilities, and had a brick station. It was also the northern terminus of the Twin Mountain & Potomac Railroad, an agricultural hauler. The canal line would basically follow the B&O Railroad out of Cumberland. At Rawlings, the canal would enter and follow the river for a while. This would happen again further east, where the canal would enter the river for over a mile. Lock 39 would be at Warrior Run in Maryland. From Cumberland to George's Creek, the cost would have been almost two and a quarter million dollars. It was estimated this option would drop the cost of coal transportation to tidewater at $1.65 per ton compared to the railroads. The then-current cost of coal transportation by the railroad was $3.26 per ton for the 212 miles required.

There was actually another option, considered by McNeill in his 1824 survey. This was to continue up the Potomac and Savage River, and reach the Youghiogheny River by means of Deep Creek. However, the Will's Creek Route was 18 miles shorter, and 440 feet lower. This survey was made in 1823 by Major J. J. Abert, Corps of Topological Engineers.

We are left to wonder what the C&O Canal system would look like today if it had been completed to Pittsburgh. Actually, you can hike or bike it to take a look.

The Great Allegheny Passage from Pittsburgh to Cumberland,

follows the path of one of the Northern option. It is a hiker-biker trail that connects Pittsburgh with Washington, DC via the C&O Canal National Historic Park at Cumberland, Maryland. The trail follows the path of the old Pittsburg & Lake Erie and Western Maryland Railroad tracks that were removed in the early 1980s. It passes through McKeesport, West Newton, Connellsville, Ohiopyle, Confluence, Rockville and Meyersdale, PA, then to Frostburg, where it follows the Western Maryland Scenic Railroad line into Cumberland. At Cumberland, the trail connects to the C&O Canal Towpath. The path is about 300 miles long, and has a maximum 2% grade, due to the use of existing canal and railroads rights-of-way. No motorized vehicles are permitted. It reaches a maximum elevation of about 2,400 feet at the Eastern Continental Divide near Deal, PA.

Westernport, MD and Piedmont WV

Westernport, Maryland, and Piedmont, West Virginia, are across the Potomac River from one another, where the George's Creek adds its flow to the bigger river.

Westernport was an easy step from Cumberland for the B&O railroad, and would have been for the canal as well. Beyond Westernport, Backbone Mountain makes things interesting. The canal had followed the water-level route of the Potomac from Georgetown to Cumberland, and could have done so to Westernport. The Savage River is a tributary of the Potomac. It is named for an 18th century surveyor. The river is now tamed by a dam built by the Army Corps of Engineers. The resultant reservoir extends some 360 acres. The dam is 184 feet high, and extends more than a thousand feet.

After the B&O reached Piedmont in 1851, the easy part was over. Now the line had to climb the 17-Mile Grade, over Backbone Mountain. This was a major engineering challenge for men and machines. The B&O line to Grafton was completed in 1852, and to Wheeling on the Ohio River in 1853. The line was open for service from Baltimore to Cincinnati and St. Louis by 1857.

The B&O's 17-Mile grade

The West End of the B&O begins at Viaduct Junction in Cumberland, and proceeds along the Potomac River Valley to Keyser and Piedmont, WV. From there, it begins the climb up the famed 17-Mile Grade to Altamont, then down the Cranberry Grade to Grafton.

The 17-Mile Grade is located on the B&O West End, west of Piedmont, WV and the paper mill at Luke, Maryland. Built in 1851 by the legendary Benjamin Latrobe, it is still used as a benchmark of engine hauling and braking performance. Coal still comes east on the line. The grade begins at the Bloomington, MD, bridge over the Potomac. This three-arch structure, referred to as Bridge 79 by the B&O, was built in 1851. From here, the 2.28% rail line snakes west up Backbone Mountain, keeping the Savage River on the right. Near the top, the line passes through the 399 foot long Hitchcock Tunnel, opened in 1865.

The halfway point of the 17-Mile Grade is at Strecker, once a watering station for the steam engines. Altamont at 2,628 feet above sea level, is the highest point on the B&O system. At Terra Alta, elevation 2,557 feet, the 17-Mile Grade gives way to the 12-mile long Cranberry grade from Rowlesburg.

Pittsburgh, PA, and the Pennsylvania Canal System

The good people of Pittsburgh prepared for the arrival of the C&O canal, and its tie-in to the Pennsylvania Canal system, the Pennsylvania Main Line of Public Works. This was the term favored by the Pennsylvania Assembly in 1824 to refer to the complex system of canals and their infrastructure in the Keystone State. This included both State-owned and canals built by Private enterprise. Unfortunately, railroads began displacing canals as a means of transportation around 1850, when the C&O Canal reached Cumberland. Financing for the section to Pittsburgh was impossible to get. In 1859, the Pennsylvania State Canals were sold. The Pennsylvania Railroad operated freight service on some of the canals for a time.

The C&O Canal was headed to the Western Division Canal system at Pittsburgh, which provided access to the three rivers there, particularly the Ohio, the longtime goal. Getting the main Pennsylvania canal across the Allegheny River into Pittsburgh required an aqueduct of 1,140 feet, the longest on the Pennsylvania Main Line route. Linking to the Ohio River, the Western Division Canal also went through an 810 foot tunnel through Grant's Hill in Pittsburgh, to the Monongahela River. This is the route that canal boots from Cumberland would have taken to enter the City. The first fully-loaded freight boat went from Johnstown to Pittsburgh in 1831 on the Western division.

Western Maryland Rail Trail

The Western Maryland Rail Trail follows the canal and the path of the Western Maryland Railroad, from near the Siding Hill Creek Aqueduct and Lock 56 to Big Pool, near Fort Frederick. It passes through the town of Hancock. It is 22.5 miles long, and paved. It parallels the canal and the tow path on the Maryland side of the Potomac. The Maryland Department of Natural Resources owns the trail in Maryland. It passes through the old WM Kessler and Stickpile tunnels, and crosses the Potomac six times.

The trail is being extended to the Indigo Tunnel, but will not continue through it due to environment impact to the largest bat refuge in Maryland. The Trail will use the canal tow path as a bypass at this point. The eventual terminus is planned to be Paw Paw WV.

And, in conclusion

We have seen how the canal and the railroad reached Cumberland, site of the National Road as well. All three transportation options were built with the idea of linking the seaboard with the Ohio River. The canal and the railroad competed with each other for workers, finance, and cargo. The railroads fed coal to the canal at Cumberland. The railroad put the earlier stagecoach lines out of business, and, except for local delivery, the freight wagons as well. Although grain from the Ohio country was the reason for construction of both systems, it was coal that became the dominant cargo. Hugh amounts of coal still flow out of western Maryland and West Virginia on the train. We can see the Industrial Revolution moving westward from the seacoast to the trans-Appalachian lands. The railroad started out being horse-powered. The National Road hosted wagons and stages, but were rapidly superseded by the railroad. Today, the successor to the B&O Railroad still hauls freight over the lines surveyed in the early 19th century. One can still ride a train westward from Baltimore, via DC, through Cumberland, to Pittsburgh. The National Road has been superseded by an Interstate, which sees heavy truck traffic. President Jefferson, who authorized the construction of the National Road, would be impressed that we can reach the Ohio from tidewater in a few hours. George Washington, who walked, canoed, and rode horse back on that path would be astounded. Today, the Ohio Country is free from French influence, and the Union is intact. We just briefly mentioned air service as a method of Internal Communication. Today, moving information is as important or more important than moving goods. You can hike or bike from DC to Pittsburgh on a single path, which uses the canal tow path and railroad right-of-ways. What will it be like 100 years from now?

For Further Reading

Abert, Colonel J. J., "Report in Reference to the Canal to connect the Chesapeake and Ohio Canal with the City of Baltimore," 1838, Washington, Government Printing Office, 1874. (University of Michigan, University Library, Michigan Historical Reprint Series, ISBN 9781418197223)

Achenbach, Joel The Grand Idea George Washington's Potomac and the Race to the West, 2004, Simon & Schuster, ISBN 0-7432-6300-6.

Bauman, William, "The Inter-Relationship between the Railroads and the C&O Canal in the Development of the Canal Basin Area, Cumberland, Maryland," 2011.

Brown, Alexander Crosby "The Patowmack Canal: America's Greatest Eighteenth Century Engineering Achievement", 1962, Virginia Canals and Navigations Society, Virginia Cavalcade, Vol. 12, pp. 40-47.

Clements, Andrew The Story of the Potomac Refining Company, 2014, avail: candocanal.org/histdocs/Clemens-PRC-book.pdf

Cotton, Robert. The Chesapeake and Ohio Canal Through the Lens of Sir Robert Cotton, 1986, Windswept House Publications, ISBN 0932433170.

Cumberland Terminus, 1992, C&O Canal Historical Park. https://www.canaltrust.org/discoveryarea/cumberland-terminus/

Evans, George Heberton, Jr. The Early History of Preferred Stock in the United States, American Economic Review, Vol. 19 N. 1, March 1929.

Feldstein, Albert The Great Cumberland Floods: Disaster in the Queen City, 2009, The History Press, ISBN-1596296887.

Garrett, Wilbur, "Patowmack Canal," National Geographic, June 1987.

Goodrich, Carter. Government Promotion of American Canals and Railroads: 1800-1890, Greenwood Publishing Group, June 1960, ISBN: 0837177731.

Hadden, James, Washington's Expeditions (1753-1754) and Braddock's Expedition (1755), with a history of Tom Fausett, the slayer of General Edward Braddock, 2009, Heritage Books Inc.; Reproduction of 1910 edition, ISBN-10: 0788423371.

Hahn, Thomas F. The C&O Canal Boatmen 1892-1924, 1980, The American Canal & Transportation Center, ISBN 0933788584.

Hahn, Thomas F. Chesapeake & Ohio Canal Old Picture Album, 1979, The American Canal & Transportation Center, 3rd edition, ASIN B0006XVFXW.

Hahn, Thomas F. The Chesapeake & Ohio Canal Lock-Houses (Monograph Series (West Virginia University. Institute for the History of Technology and Industrial Archeology, May 1996, ISBN 1885907036.

Hahn, Thomas F. George Washington's Canal at Great Falls, Va., American Canal & Transportation Center, June 1976, ISBN: 093378855X.

Hahn, Thomas F. C. and O. Canal, an Illustrated History, American Canal & Transportation Center, June 1981, ISBN: 0933788592.

Hahn, Thomas F. Cement Mills along the Potomac River, West Virginia University Press, 1994, ISBN 1885907001.

Hahn, Thomas F. Towpath Guide to the C & O Canal (Georgetown, Seneca, Harper's Ferry, Fort Frederick,

Cumberland), American Canal & Transportation Center, ISBN 093378872X.

Hansrote, Hazel "Cumberland, Maryland Terminus of the C&O Canal, 1850-1924: Scrapbook," 1980, Preservation Society of Allegany County, ASIN B0006XVFXW.

Harlow, Alvin F. When Horses Pulled Boats, American Canal & Transportation Center, April 1983, ISBN-0933788436.

Heine, Cornelius W. "The Washington City Canal," Records of the Columbia Historical Society, Vol. 53/56, 1953-56, pp. 1-27.

High, Mike. The C&O Canal Companion, Johns Hopkins University Press, April 1997, ISBN: 0801855705.

"Steam on the C&O Canal," Hunt's Merchant's Magazine, June 6, 1858,

Ierley, Merritt. Traveling the National Road, 1990, Woodstock, NY: Overlook Press, ISBN 0-87951-394-2.

Kapsch, Robert J. Ph.D. "Baltimore and the Maryland Cross-Cut Canal: 1820-1851, ASCE, 2005. Baltimore Civil Engineering History Conference, p. 117-155.

Kapsch, Robert J. The Potomac Canal George Washington and the Waterway West, 2007, West Virginia University Press, ISBN 9781-933202-18-1.

Kytle, Elizabeth. Home on the Canal, Johns Hopkins Univ. Press, April 1996, ISBN: 0801853281.

Leyland, Herbert The Ohio Company, A Colonial Corporation, Nabu Press, ISBN 781176894211, Quarterly Publication of the Historical & Philosophical Society of Ohio, Volume XVI, 1921 No. 1, Jan-Jun.

Martineau, John "Frederick County Canal, Report of the Engineer employed, on behalf of the citizens of Frederick-Town, to make a survey and furnish estimates for a Canal from said town to the Chesapeake and Ohio Canal," holdings of the Historical Society of Frederick County, Frederick-Town Herald, 1830.

Mould, David H. Dividing Lines: Canals, Railroads and Urban Rivalry in Ohio's Hocking Valley, 1825-1875, Wright State University Press, March 1994, ISBN: 1882090063.

Peck, Garrett, The Potomac River: A History and Guide, The History Press, March 6, 2012, ISBN-10: 1609496000.

Pickell, John, A New Chapter in the Early Life of Washington, in Connection with the Narrative History of the Potomac Company, Scholarly Publishing Office, University of Michigan Library, 2005, ISBN 1425514146.

Riley, Harvey The Mule: A Treatise on the Breeding; Training; and Uses to Which He May be Put, 2007 reprint, Bibliobazaar, ISBN 1434695441.

Sabatke, Mark Discovering the C&O Canal, 2003, Schreiber, Sehngold Publishing, ISBN 1887563679.

Sanderlin, Walter S. Great National Project : A History of the Chesapeake and Ohio Canal, 1946. reprinted 2005, Eastern National Park and Monument Association, ISBN 9781590910498.

Shank, William H. Amazing Pennsylvania Canals, 4th edition, American Canal & Transportation Center, June 1981, ISBN: 0933788371.

Shaughnessy, Jim. Delaware & Hudson: The History of an Important Railroad Whose Antecedent Was a Canal Network to Transport Coal, Syracuse University Press (Trade) Publication,

February 1997, ISBN: 0815604556.

Shaw, Robert E. Canals for a Nation : The Canal Era in the United States,1790-1860, University Press of Kentucky, February 1993. ISBN: 0813108152;

Stakem, Patrick H. "Coal to the Western Terminus, Canal-Railroad Connections in Cumberland, Maryland," Sept. 1995, On the Towpath, Vol. XXVII, no. 3 p. 10-11.

Stakem, Patrick H. "The C&O Canal and the Railroads, Cooperation, Synergy, and Competition," January 1993, invited paper, presentation to a joint meeting of the C&O Canal Historical Society and the Western Maryland Chapter, National Railway Historical Society, Western Maryland Station Center, Cumberland, MD, reprinted in June 1993 issue of On the Towpath, the newsletter of the C&O Canal Historical Society.

Stakem, Patrick H. The History of the Industrial Revolution in Western Maryland, 2010, PRB Publishing, ASIN B004LXOJB2.

Trowbridge, Prof. W. P. Reports on the Water-Power of the United States, Bureau of the Census, Government Printing Office, 1885, www2.census.gov/prod2/decennial/documents/1880a_v16-01.pdf

Unrau, Harlan D. "Historical Research Study, Chesapeake & Ohio Canal," United States Department of Interior, National Park Service, Chesapeake & Ohio Canal National Historical Park , Hagerstown, Maryland , August 2007.

Warren-Findley, Jannelle "A History of the C&O Turning Basins at Cumberland, Maryland 1835-1958," C&O Canal Historical Park, NPS, 12/11/07.

Way, Peter. Common Labor : Workers and the Digging of North American Canals 1780-1860, Johns Hopkins University Press, November 1996, ISBN: 0801855225.

Wolfe, George Hooper, I Drove Mules on the C&O Canal,1969, Dover Graphic Associates, ASIN B00137CHNW.

Young, Rogers W. The Chesapeake and Ohio Canal and the Antebellum Commerce of Georgetown, January 1940, NPS.

Youatt, William and Skinner, John S. The Horse: Together with a General History of the Horse and an Essay on the Ass and the Mule, 2010 reprint, Kessinger Publications, ISBN 1163797754.

Youatt, William The Horse, with a Treatise of Draught (by Isambard Kingdom Brunel), 1831, Cornell University Library, reprint 2009, ISBN 1112359052.

"Draft, Executive Summary, for Phase I, Archeological Excavation in Area of Proposed Turning Basin, Crescent Lawn Architectural District, (18AG227), City of Cumberland, Allegany County, Maryland," J. Milner Associates, May 2000, for the U. S. Corps of Engineers, Baltimore Division.

"A Profile of the Frederick Town Canal" drawings and plans of the proposed extension of the C&O Canal along the Monocacy River c. 1830, Maryland State Archives (MSA SC 1386-12 – MSA SC 1386-20).

Surveys for the Extension of the Chesapeake and Ohio Canal, United States Engineer Office, Cincinnati, Ohio, February 10, 1876, in Annual Report of the Chief of Engineers to the Secretary of War, for the Year 1876, Part I & Part II, Washington: Government Printing Office, 1876. 44th Congress, 2d Session, U.S. House of Representatives.

"The Economic Impact of the C&O Canal on Canal Communities in Washington County, Maryland," http://www.whilbr.org/CandOCanal/index.aspx.
Wikipedia, various.

Places

Braack, Craig R. Allegany County in the Twentieth Century: Stories of Change, Allegany county Bi-centennial History, Donning Company Publishers, January 2005, ISBN-10 1578643341.

Gude, Gilbert Small Town Destiny: The Story of 5 Small Towns Along the Potomac Valley (Brunswick, Williamsport, Hancock, Maryland; Shepherdstown, Paw-Paw, West Virginia), Lomond Publications June 1989, ISBN 0912338695.

Kaminkow, Marion J. Maryland A to Z; A Topographical Dictionary, 1985, Baltimore: Magna Carta Book Co.

Mulligan, Kate, Towns along the Towpath, 1996, Washington, DC: Wakefield Press, ISBN 0-9655552-0-8

Scharf, J. Thomas, History of Western Maryland, Vol. 2, p. 1488-1489. ISBN 0832838764.

Whetzel, Dan, Allegany County, Arcadia Press, Images of America Series, 2011, ISBN 0738587044.

Brunswick, MD

Rubin, Mary H. Brunswick (MD), Arcadia Press, Images of America Series, 2007, ISBN 0738552739.

Cumberland, MD

Feldstein, Albert L. Downtown Cumberland 1950-1980, 1994, Cumberland, MD: Commercial Press.

Feldstein, Albert L. Feldstein's Historic Coal Mining and Railroads of Allegany County, Maryland, 2000, ISBN 097016050X.
Feldstein, Albert L. Tour Guide to Historic Sites in Allegany

County, Maryland, 1990, Cumberland, MD Commercial Press.

Feldstein, Albert L. Allegany County, Arcadia Press, Images of America Series, 2006, ISBN 9780738543819.

Hunt, J. William. The Story of Cumberland, Maryland, 1965, Allegany County Historical Society, Cumberland, MD

Lowdermilk, W. H. History of Cumberland, MD, 1878, Washington, DC, reprinted, Regional Publishing Co., 1976, Baltimore, ISBN 0-8063-7983-9.

"Queen City Story as Told in News Articles," Cumberland, MD, Preservation Society of Allegany County, 1981.

Schwartz, Lee G.; Feldstein, Albert; Baldwin, Joan H., Allegany County, A Pictorial History, 1980, Virginia Beach, VA: The Donning Co., ISBN-0898650178.

Stakem, Patrick H. Cumberland, Then & Now, Arcadia Press, 2011, ISBN 9780738586984.

Thomas, LL.D., James W., and Williams, Judge T. J. C. History of Allegany County, Maryland, 1923, reprinted, Wildside Press, 2011, ISBN-1434426297, and -1434426300.

(see also, Long, Helen, Index to Scharf's History of Western Maryland, Volume I & II, 2013, ISBN-0806345667

Stegmaier, Harry, Jr., et al. Allegany County - A History, 1976, Parsons, WV, McClain Publishing, ISBN 0870122576.

Weaver, Joseph H. Cumberland 1787-1987 A Bicentennial History, 1987, Cumberland, MD: Precision Printing Co. ASIN-B0007165K6.

Zumbrun, Champ A History of Green Ridge State Forest (MD)

2010, The History Press, ISBN 1596299029.

Eckhart

Hughes, George Wurtz. "Extracts from reports of an examination of the coal measures belonging to the Maryland mining company, in Allegany county; and of a survey for railroad from the mines to the Chesapeake and Ohio canal, at Cumberland," 1837, Printed by Gales and Seaton, Washington (available: Pratt).

Report of President & Board of Directors of Cumberland Coal & Iron Company to the Stockholders, Feb. 11, 1853, New York, John F. Trow, Printer. Avail:
https://catalog.hathitrust.org/Record/100254666

Rankin, Robert G. "Report on Cumberland Coal Basin," 1855, New York, John F. Trow, Printer.

Stakem, Patrick H. Eckhart Mines, the National Road, and the Eckhart Railroad, 2011, ASIN B004KSQVWO.

Frederick and the Monacacy River

http://www.dnr.maryland.gov/boating/mdwatertrails/pdfs/monocacywatertrailmap.pdf

Georgetown

Gorr, Louis F. "The Foxall-Columbia Foundry: An Early Defense Contractor in Georgetown," Records of the Columbia Historical Society, Washington, D.C., Vol. 71/72, The 48th separately bound book, 1971-72, pp.34-59. www.jstor.org/stable/40067769.

Harpers Ferry, WV

Adelman, Garry E. and Richter, John R. 99 Historical Images of Harper's Ferry, 2007, Center for Civil War Photography, ISBN

0978550848.

Gilbert, David T. Waterpower Mills, Factories, Machines, and Floods at Harper's Ferry, WV 1762-1991, Harpers Ferry Historical Association; 1999, First Edition, ISBN-10: 0967403308

Johnson, Mary, "A Nineteenth Century Mill Village: Virginius Island 1810-60," West Virginia History, Vol. 54, 1995, pp. 1-27.

Nasby, Dolly Harper's Ferry, Then & Now, Arcadia Press, 2007, ISBN 9780738544144.

Smith, Merritt Roe Harper's Ferry Armory and New Technology, 1980, Cornell University Press, ISBN 0801491819.

Lonaconing

Alexander, John H. "George's Creek Coal and Iron Company," 1836. [Baltimore?, 1837], available: Frostburg University library: Special Collections, TN805.Z6G3, also Pratt Library, Baltimore, D9549.G4A3q.

Harvey, Katherine A. "The Lonaconing Journals: The Founding of a Coal and Iron Community 1837-1840," 1977, Transactions of the American Philosophical Society, Philadelphia, Vol. 67, Part 2, March, 1977.

Harvey, Katherine A. "Building a Frontier Ironworks: Problems of Transport and Supply, 1837-1840," Maryland Historical Magazine, Vol. 70, No. 2 Summer 1975.

Harvey, Katherine A. The Best-Dressed Miners - Life and Labor in the Maryland Coal Region 1835-1910, 1969, Cornell University Press, ISBN 0801404940.

Lonaconing - Home in the Hills, 1986, Lonaconing, MD: George's Creek Promotion Council, ASIN B000HZXQ04.

Stakem, Patrick H. Lonaconing Residency: Iron Technology and the Railroads, PRB Publishing, 2011, ASIN B004L62DNQ.

Mt. Savage, MD

Aldridge, Howard Redford. "The Mount Savage Iron Works," 1924, Records of Phi Mu Fraternity, University of Maryland at College Park Libraries.

Allen, Jay Douglas. "The Mount Savage Iron Works, Mount Savage, Maryland a case study in pre-Civil War industrial development," 1970, Thesis (M.A.) - University of Maryland, 1970.

Bryant, William Cullen. "Mount Savage, 1860," Saturday Evening Post, 1860, Reprinted in Tableland Trails, Vol. I, No. 3, Fall, 1953, Oakland, MD

Carney, Charles. "The History of Mount Savage," May, 1967, Project 67-014-005, Cooperative Extensive Service, University of Maryland.

Deffenbaugh, Mrs. Roy. "History of Mount Savage, Maryland," Mount Savage High School, 1968.

Minor, D. K. (ed.). Tour of Mount Savage, American Railroad Journal, Summer, 1844.

Stakem, Patrick H. "Mt. Savage Locomotive Shops," National Railway Historical Society Bulletin, Spring/Summer 1997.

Ridgeley, WV

Clites, Sr. Gary Ridgeley and Carpendale, West Virginia From 1750 A History, 2008, Knobley Mountain Press, ISBN 978-1-4357-2044-2.

Washington, DC

L'Enfant, Pierre Charles, A Plan Wholly New, Library of Congress, 1993, ISBN 0-8444-0699-6.

Williamsport, MD

Rubin, Mary H. Williamsport, Arcadia Press, Images of America Series, 2005, ISBN 9780738541761.

Wolfe, George Harper The Town of Williamsport: A Historic Memorabilia, 1997 (out-of-print), ASIN B0006XZJ8Y.

Railroads

Kelly, Jacques, Trackside Maryland: From Railyard to Main Line, Johns Hopkins Press, 2003, ISBN 0801873231.

"A Railfan's Guide to the Cumberland Area," pamphlet, Western Maryland Chapter of the National Railway Historical Society, Inc. Cumberland, MD.

Stakem, Patrick H. Railroading around Cumberland, Arcadia Press, 2008, ISBN 9780738553658.

Amtrak

Soloman, Brian Amtrak, 2004, Motorbooks International, ISBN 0760317658.

Staff of Amtrak, Amtrak, an American Story, 2011, Kalmback Books, ISBN 0871164442.

Baltimore & Ohio Railroad

Avery, Carlos P. E. Francis Baldwin, Architecture, the B&O,

Baltimore, & Beyond, 2003, Baltimore Architectural Foundation, ISBN-0-97297430-X.

Dilts, James D. The Great road, the Building of the Baltimore and Ohio, the Nation's First Railroad, 1828-1853, 1993, Stanford University Press, ISBN-0804722358.

Edson, William D. Steam Locomotives of the Baltimore & Ohio An All Time Roster, 1992, ISBN 0-9632913-0-0.

Harwood, Jr., Herbert H. (1979). Impossible Challenge: The Baltimore & Ohio Railroad in Maryland. Baltimore, MD: Barnard, Roberts. ISBN 0-934118-17-5.

Henderson, John. Blue Diesels & Black Diamonds, 1991, H&M Productions, ISBN-0962903752.

Hollis, Jeffrey R. and Roberts, Charles S. East End B&O's 'Neck of the Bottle', 1992, Bernard, Roberts & Co. ISBN 0-934118-19-1.

Roberts, Charles S. West End, Cumberland to Grafton 1848-1991, 1991, Barnard, Roberts & Co., Baltimore, MD, ISBN 0934118183.

Roberts, Charles S. and Hollis, Jeffrey R. East End, B&O's Neck of the Bottle, Harpers Ferry to Cumberland 1842-1992, Barnard, Roberts & Co., Baltimore, MD, ISBN 0-9341118-19-1.

Roberts, Charles S. Sand Patch, Clash of Titans, Cumberland to Connellsville and branches 1837-1993 Barnard, Roberts & Co., Baltimore, MD, ISBN 0-934118-20-5.

Jacobs, Timothy (ed.) The History of the B&O America's First Railroad, 1989, Brompton Books Corp. ISBN-0517676036 .

Welsh, Joe, Baltimore & Ohio's Capitol Limited and National Limited (Great Passenger Trains), Voyageur Press; 1st edition, 2007, ISBN-10: 0760325332.

Kirsch, Richard. "Rails to Cumberland," 1/5/92-11/1/92, Cumberland Sunday Times News, in 10 parts.

Mellander, Deane. B&O Thunder in the Alleghenies, 1993, Carstens, ISBN 0911868453.

Newell, Dianne The Failure to Preserve the Queen City Hotel, Cumberland, Maryland, Washington: Preservation Press, National Trust for Historic Preservation in the United States, 1979, c1975, ISBN 0-89133-023-2.

Nixon, David William. "Queen City Station-Hotel," Blue Print Series, 1992.

Stover, John F. History of the Baltimore & Ohio Railroad, 1987, Purdue University Press, ISBN 0-911198-81-4.

Stegmaier, Harry I. Jr. B&O Passenger Service 1945-1971 Volume I: The Route of the National Limited, 1993, TLC Publications, ISBN-188308900X.

Stegmaier, Harry I. Jr. B&O Passenger Service 1945-1971 Volume 2: The Route of the Capital Limited, 1993, TLC Publications, ISBN-1883089212 .

Summers, Festus P. The B&O in the Civil War, 1939, (reprinted 1993), Stan Clark Military Books, ISBN 1-879664-14-3

Chessie and CSXT

Biery, Thomas A. Chessie System, Cumberland Action, 1999, Railroad Press, ISBN 0-9657709-3-1.

Dixon, Jr., Thomas W. The Chessie Era, 1990, TLC Publishing, ISBN 0-9622003-2-8.

Ori, Dave (2006). Chessie System (MBI Railroad Color History). Voyageur Press. ISBN 978-0760323397.

Nuckles, Douglas B. and Dixon, Jr., Thomas W. Diesel Locomotives of CSXT and Predecessors in Color, 1993, TLC Publishing Inc, ISBN-1883089042.

Chessie System Historical Society, http://www.trainweb.org/CSHS/

Conrail

Doherty, Tim Conrail, 2004, Motorbooks International, ISBN 076031425X.

CSX

Soloman, Brian, CSX, 2005, Motorbooks International, ISBN 0760317968.

Stakem, Patrick E. and Stakem, Patrick H. CSX Diesel Locomotives in Color, 2000, Motorbooks International, ISBN 1883089433.

Cumberland Valley Railroad

Westhaeffer, Paul J. History of the Cumberland Valley Railroad, 1835-1919, 1979, Washington Chapter, NRHS, Inc. ISBN 093395400X.

http://d_cathell.tripod.com/cvrrmain.html

Historic Railroad Lines

Arnold, Newt, "The T. M. &P. - Two Mules and a Pony," April 16, 1980, Keyser (WV) News-Tribune, reprinted in the Automatic Block, Nov-Dec 1984 (newsletter of the Western Maryland

Chapter, NRHS, Inc. Cumberland, MD)

Hicks, H. Ray. "The Cumberland and Pennsylvania Railroad," Railroad & Locomotive Historical Society Bulletin, No. 66, March, 1945, pp. 36-50.

Hicks, H. Ray. " The George's Creek and Cumberland Railroad," Railroad & Locomotive Historical Society Bulletin, No. 85, March, 1952.

Hicks, H. Ray. "The Pennsylvania Railroad in Maryland,", No. 85, Railroad & Locomotive Historical Society Bulletin, March, 1952.

Lardner, Dionysius The Steam Engine Explained and Illustrated (7th Edition) With an Account of its Invention and Progressive Improvement, and its Application to Navigation and Railways; Including also a Memoir of Watt, 1840, ASIN-B00CKCKO7U.

Mellander, Deane. Rails to the Big Vein, the Short Lines of Allegany County, Maryland, January, 1981, Potomac Chapter, NRHS, Inc.

Mellander, Deane. Cumberland and Pennsylvania Railroad, 1981, Newton, NJ: Carstens Publishers, Inc., ISBN 911868-42-9.

Metcalf, Paul Waters of Potomack, North Point Press, 1982, ISBN-0-86547-090-1.

Stakem, Patrick H. "The Engines Cumberland," The Automatic Block, (newsletter of the Western Maryland Chapter, NRHS, Inc., Cumberland, MD, March, 1992, Vol. 14, no. 3.

Stakem, Patrick H. "The George's Creek Railroad 1853-1863," Nov. 1995, The Automatic Block, Vol. 17, No. 11.

Stakem, Patrick H. "The Eckhart Branch Railroad, 1846-1870," Jan. 1996, The Automatic Block, Vol. 18, No. 1.

Stakem, Patrick H. "The Mount Savage Rail Road 1845-1854," June 1995, The Automatic Block, Vol. 17, No. 6; reprinted in Cumberland Times, Sept. 30, 1995, Railfest special section.

Stakem, Patrick H. "The Potomac Wharf Branch," Sept. 1995, Baltimore, MD: Bullsheet.

Stakem, Patrick H. "The Earliest Railroad Activities in Western Maryland, 1828-1870," 1996, J. Alleghenies, Vol. XXXII, ISSN-0276-7449.

Stakem, Patrick H. "A Visit to the Kesler Tunnel," Nov. 1991, The Automatic Block, Vol. 13, no. 11.

Stakem, Patrick H. "Engines of the Eckhart Branch," Cumberland Times News, Sept. 30, 1996, Railfest special section.

Stakem, Patrick H. The C&P Railroad Revisited, PRRB Publishing, 2002, ISBN 0972596607.

Stakem, Patrick H. T. H. Paul and J. A. Millholland, Master Locomotive Builders of Western Maryland, 2001, ASIN B004LGT00U.

Urbas, Anton. "GC&CRR Passenger Train Service," 1993, J. Alleghenies, Vol. XXIX-1993, pp. 38-47.

92 Years of Transportation Progress by the C&P Railroad and its Contribution Towards the Development of Cumberland, and Allegany County, Maryland, C&P Railroad, 1937, Cumberland, MD.

Rail Tracks in Allegany County, Maryland, Book 1, 1980, Preservation Society of Allegany County, MD, Cumberland, MD.

Wallace, Mark R. The State Line Branch, BMX, Vol. 41, n 1,

Spring 2012. (BMX or Blue Mountain Express is the bulletin of the Western Maryland Railway Historical Society, Union Bridge, MD).

MARC

"MARC," Sept. 90, Passenger Train Journal

"MARC - Growth Industry," Oct. 1994, Passenger Train Journal, issue 202

Schneider, Paul D. "On the MARC," Jan. 1994, Trains, Kalmbach, Vol. 54 no 1 pp. 51-56.

Stakem, Patrick H. MARC: Maryland Area Rail Commuter, a Rider's Guide, 2011, PRRB Publishing, ASIN B004U7FKQS.

Norfolk & Western; Norfolk Southern

Borkowski, Richard C. Norfolk Southern Railway, 2008 Voyageur Press, ISBN 0760332495.

Harris, Nelson Norfolk and Western Railway (VA), 2003, Arcadia Press, 0738515272.

Short lines & Logging Railroads

Kline, Benjamin F.G., Jr. Tall Pines and Winding Rivers The Logging Railroads of Maryland, 1976. (Reprinted by the Western Maryland Chapter, National Railway Historical Society in Cumberland.

Traction

King, LeRoy O. 100 Years of Capital Traction, Dallas, TX: Taylor 1972. (This book is the definitive reference to the trolleys of the Washington, DC area. It is available at the National Capital Trolley

Museum (NCTM) in Kensington, MD).

"Street Cars in the Nation's Capital," NCTM Journal, Vol. 11, No. 2, August 1995.

Tosh, Francis B. "Cumberland & Westernport Electric Railway," 1963, Bulletin of the National Railway Historical Society, Vol. 28 No. 4 p. 24.

Western Maryland Railroad

Biery, Thomas A. Western Maryland, the Connellsville Extension, May/June 1995, The Railroad Press, Issue 24.

Cook, Roger and Zimmerman, Karl. Western Maryland Railway - Fireballs and Black Diamonds, 1981, Howell-North Books, San Diego, Ca., ISBN 0-8310-7139-7.

Grenard, Ross and Krause, John. Steam in the Alleghenies, Western Maryland, 1981, Newton, NJ: Carstens Publishing Co., ASIN B002J02I8G.

Hicks, H. Ray. "The Connellsville Extension of the Western Maryland Railway," Railroad & Locomotive Historical Society Bulletin, No. 85, March, 1952.

Pennypacker, Bert. To Cumberland & Beyond, NRHS Bulletin, Vol. 54, no. 6, 1989, reprinted 1995 by the Western Maryland Chapter, NRHS, Inc. (Cumberland, MD).

Price, William. Western Maryland Steam Album, 1986, Potomac Chapter, NRHS, Inc., Kensington, MD, ASIN B000TXM7MC.

Salamon, Stephen J. & Hopkins, William E. The Western Maryland Railway in the Diesel Era, 1991, Silver Spring, MD: Old Line Graphics, ISBN 187931407X.

Stakem, Patrick E. and Stakem, Patrick H. Western Maryland Diesel Locomotives, 1996, TLC Publishing, ISBN 1883089247.

Williams, Harold A. The Western Maryland Railway Story: A Chronicle of the First Century - 1852-1952, Baltimore: Western Maryland Railway Co. 1952.

"Western Maryland's Future. Baltimore City's Holdings Formally Transferred to the Fuller Syndicate," New York Times June 28, 1902.

Western Maryland Scenic Railroad

Kraemer, Thomas K. Western Maryland Scenic Railroad, 2003, RRTrax, ISBN 0974306002.

Kirsch, Richard, and Rundle, John, "Scenic Western Maryland Railroad Travel Guide," 1990, Western Maryland Chapter, NRHS, Inc.

Stakem, Patrick H. "A Railfan's Guide to the Allegany Central," privately published for Western Maryland Railway Historical Society's Cumberland Convention, 1989.

"Western Maryland Weekend", Railpace News Magazine, August, 1991.

T. H. Paul

Best, G. M. "Thomas H. Paul & Son, Locomotive Builders," Railway & Locomotive Historical Society Bulletin , No.141, Autumn, 1979, pp. 19-26.

Stakem, Patrick H. "T.H. Paul, Master Locomotive Builder of Frostburg, MD" J. Alleghenies, Vol. XXXIII, 1997.

Stakem, Patrick H. T. H. Paul and J. A. Millholland, Master

Locomotive builders of Western Maryland, 2011, PRRB Publishing, ASIN-B004LGTOQU.

West Virginia Central & Pittsburg

Hicks, H. Ray. "The West Virginia Central & Pittsburg," Locomotive & Railway Historical Society, Bulletin #113, 1969.

West Virginia Central & Pittsburg Railway Company, 1899, reprint 1992, Parsons, WV: McClain Printing Co., ISBN 87012-418-8.

Clark, Alan The West Virginia Central and Pittsburg Railway, 2003, TLC Publishing, ISBN 1-883089-87-5.

National Road

Remarks on the Intercourse of Baltimore with the Western Country - With a View On the Communications Proposed Between the Atlantic and Western States, Published by Joseph Robinson, Baltimore, 1818, available on books.google.com.

Hill, Harry S. The Conestoga Wagon, A Short Story of a Ship of Inland Commerce, 1930, Phillips & Godshalk, ASIN B00087XJ4Y.

Ierley, Merritt Traveling the National Road: Across the Centuries on America's First Highway, 1993, Overlook, ISBN 9780879514952.

Omwake, John Conestoga Six Horse Bell Teams, 1930, ASIN B001OFPJHY.

Meyer, Balthasar Henry History of Transportation in the United States before 1860, Carnegie Institute of Washington, 1917, republished 1948, ASIN-B003RCEGZI.

Peyton, Billy Joe "To Make the Crooked Ways Straight and the

Rough Ways Smooth" The Federal Government's Role in Laying out and Building the Cumberland Road," Dissertation, Eberly College of Arts and Sciences, WVA, 1999.

Raitz, Karl B. et al, The National Road, JHU Press, 1996, ISBN 0801851564.

Reist, Arthur L. Conestoga Wagon – Masterpiece of the Blacksmith, Lancaster, PA; 1st Edition, 5th Printing edition 1975, ASIN B000OL08V6.

Peyton, Billy Joe "To Make the Crooked Ways Straight and the Rough Ways Smooth" The Federal Government's Role in Laying out and Building the Cumberland Road," Dissertation, Eberly College of Arts and Sciences, WVA, 1999.

Shumway, George, Durel, Edward, and Frey, Howard C. Conestoga Wagon, 1750-1850, 1964, Early American Industries Association, Inc, ASIN B000OL21Z2.

http://stagecoachfreightwagon.org/

http://www.colonialsense.com/Society-Lifestyle/Signs_of_the_Times/Conestoga_Wagon.php

http://www.scribd.com/doc/46877173/The-Development-of-Transportation-Facilities-as-a-Stimulus-to-Growth-in-Baltimore-1800-1853

Forrest, Earle R. "National Pike, Road of History, Romance," The Washington Reporter, Washington, PA, March 18, 1955, P. 8, http://www.ourfamilyhistories.com/hsdurbin/pike/part2.html

http://marylandnationalroad.org/history-of-the-road/

Misc topics

Bauman, William "Hustle & Bustle on the C&O Canal - 1851," 2011, from the Cumberland Alleganian, 1851, On the Towpath, June 2011, Vol. XLIII, No. 3,

Beachley, Charles E. History of the Consolidated Coal Company 1864-1934, 1934, Consolidation Coal Company, New York.

Darlington William M. Christopher Gist's Journals; With Historical, Geographical and Ethnological Notes and Biographies of His Contemporaries, Kessinger Publishing, LLC, September 10, 2010, ISBN- 1163446092.

Dunbar, Seymour, A History of Travel in America, Showing the Development of Travel and Transportation From the Crude Methods of the Canoe and the Dog-Sled to the Highly Organized Railway Systems of the Present..., reprinted, General Books LLC (2010), ISBN 1152314432.

Earle, Thomas, Treatise on Rail-Roads and Internal Communications, Nabu Press (March 16, 2010), ISBN- 1147258511.

Hulbert, Archer B. The Paths of Inland Commerce, 1920, Yale University Press (Echo Library 2009), ISBN 1406850624.

Hulbert, Archer B. Washington and the West; being George Washington's diary of September, 1784, kept during his journey into the Ohio basin in the interest of a commercial union between the Great Lakes and the Potomac River, Nabu Press (August 9, 2010), ISBN 1177081121.

Lacoste, Kenneth C., Wall, Robert D. An Architectural Study of the Western Maryland Coal Region: The Historic Resources, Maryland Geological Survey, 1989.

Metzger, Bill, The Great Allegheny Passage Companion: Guide to History & Heritage Along the Trail, 2003, Local History Co., ISBN

097118352X.

Muller, Edward K. (ed.) An Uncommon Passage: Traveling through History on the Great Allegheny Passage Trail, 2009, University of Pittsburgh Press, ISBN 9780822943662.

Pred, Allan R. *Urban Growth and the Circulation of Information 1790-1840*, Harvard University Press, 1973.

Randolph, B.S. "History of the Maryland Coal Region," Journal of the Alleghenies, Vol. XXIX-1993, pp. 47-62.

Shaw, Mary and Weil, Roy, Linking Up: Planning Your Traffic-Free Bike Trip Between Pittsburgh, PA and Washington, DC - 3rd Edition, 2007, Great Allegheny Press, ISBN 9780979210815.

Stanton, Richard. Potomac Journey, Fairfax Stone to Tidewater, Smithsonian Press, 1996, ISBN 1560986603.

Toomey, Daniel Carroll. The Civil War in Maryland, 1983, Baltimore: Toomey Press, ISBN 0961267003.

Ware, Donna M. Green Glades and Sooty Gob Piles, Maryland Historical Trust, 1991, ISBN 1878399012.

Wilner, Alan M. Maryland Board of Public Works - a History, 1984, Maryland Hall of Records, Annapolis, MD.

Fourteenth Annual Report of the Bureau of Statistics and Information of Maryland. 1905.

Malone, Patrick M. Waterpower in Lowell: Engineering and Industry in Nineteenth-Century America, Johns Hopkins Introductory Studies in the History of Technology, 2009, ISBN-0801893062.

Notes

Automatic Block is the Newsletter of the Western Maryland Chapter, National Railway Historical Society, P. O. Box 1331, Cumberland, Maryland, 21501.

On the Towpath is the newsletter of the C&O Canal Historical Society, Glen Echo, MD.

Bullsheet was published by Allen Brougham of Sykesville, MD, and is now a website.

BMX, or Blue Mountain Express, is the bulletin of the Western Maryland Railway Historical Society, Union Bridge, MD.

WHILBR is the Western Maryland's Historical Library, at http://www.whilbr.org .

Videos

1. *Feldsteins Historic Allegany County A Video Postcard Extravaganza*, 1990, Albert L Feldstein, Commercial Video Services, Cumberland, MD

2. *Allegany Rails*, Vol. 1, (B&O), and Vol. 2, (Western Maryland), William Price.

3. *Western Maryland Scenic Railroad*, 1991, Pentrex.

4. *Into the Allegheny Range*, 4 Volumes, 7 tape set, Iron Horse America Productions, 1995, Pentrex.

5. *MARC,* Iron Horse Video, 1994.

Websites

The C&O Canal Association, http://www.cand0canal.org

The City of Cumberland, http://www.ci.cumberland.MDus/

The National Park Service, www.nps.gov

6. National Capital Trolley Museum, http://www.dctrolley.org/

City of Hancock site at: http://www.hancockMDcom/

City of Brunswick: http://www.bhs.edu/brun/home.html

Western Maryland Scenic Railroad site is at: http://www.wmsr.com/

Harper's Ferry: http://www.nps.gov/hafe/home.htm

Railroad sites at:
    www.csxt.com
    www.nscorp.com
    www.amtrak.com
    www.conrail.com
www.infor1.
http://www.fred.net/kathy/canal.htmlm.uMDedu:8080/UMS+State/MD_Resources/MDOT/mta/services/marc.html

National Road http://www.cumberlandroadproject.com/

WesternMarylandRailTrail
http://www.westernmarylandrailtrail.org/WMRT/

If you enjoyed this book, you may enjoy some of my other titles.

Stakem, Patrick H. Cumberland & Pennsylvania Railroad Revisited, 2011, PRRB Publishing, not currently in print

Stakem, Patrick H. Eckhart Mines, The National Road, and the Eckhart Railroad, 2011, PRRB Publishing, ASIN B004KSQVWO.

Stakem, Patrick H. The History of the Industrial Revolution in Western Maryland, 2011, PRRB Publishing, ASIN B004LX0JB2.

Stakem, Patrick H. Down the 'crick: the Georges Creek Valley of Western Maryland, 2014, PRRB Publishing, ASIN B00LDT94UY.

Stakem, Patrick H. Lonaconing Residency, Iron Technology & the Railroad, 2011, PRRB Publishing, ASIN B004L62DNQ.

Stakem, Patrick H. T. H. Paul & J. A. Millhollland: Master Locomotive Builders of Western Maryland, 2011, PRRB Publishing, ASIN B004LGT00U.

Stakem, Patrick H. Tracks along the Ditch, Relationships between the C&O Canal and the Railroads, 2012, PRRB Publishing, ASIN B008LB6VKI.

Stakem, Patrick H. From the Iron Horse's Mouth: an Updated Roster from Ross Winans' Memorandum of Engines, 2011, PRRB Publishing, ASIN B005GM4012.

Stakem, Patrick H. Iron Manufacturing in 19th Century Western Maryland, 2015, PRRB Publishing, ASIN B00SNM5EIU.

Stakem, Patrick H. Railroading around Cumberland, 2012, Arcadia Press, ISBN- 0738553654.

Stakem, Patrick H. Cumberland (Then and Now), 2012, Arcadia Press, ISBN-0738586986 , ASIN B009460QNM.

Stakem, Patrick H. Fort Cumberland, Global War in the Appalachians: a Resource Guide, 2012, PRRB Publishing, ASIN- B0088BWK06.

Stakem, Patrick H. Ross Winans, an ingenious mechanic of Baltimore, 2017, PRRB Publishing, ASIN- 1520207298.

Stakem, Patrick H. Mount Savage, Iron Empire, 2016, ISBN-978-1549650413.

Stakem, Patrick H. Studebaker Wagons, 2018, ISBN-978-109146490 .

Stakem, Patrick H. Riverine Ironclads, Submarines, and Aircraft Carriers of the American Civil War, 2019, ISBN-978-1089379287.

Stakem, Patrick H. The Snowdens' Iron Works, 2019, ISBN-978-1070945699 .

Stakem, Patrick H. Transportation Options on the Frontier, 2019, ISBN-978-1091059481.

Stakem, Patrick H. Savage Factory, Cotton to Canvas, by Water & Steam, 2019, ISBN-978-1731437983.

www.ingramcontent.com/pod-product-compliance
Lightning Source LLC
Chambersburg PA
CBHW031628210526
45464CB00004B/1805